Endangered Species

Sower Grounds

The Rock

Herm

Pinockle

North Grounds

speckled Apron

Oasi

& Clarks

Little Ledge

PLUM ISLAND

PLUM ISLAND SOUND

Castle Neck

IPSWICH BAY

ESSEX BAY

Endangered Species

Chronicles of the Life of a New England Fisherman and the F/V *Ellen Diane*

David Goethel

Peter E. Randall Publisher
Portsmouth, New Hampshire
2023

Peter E. Randall Publisher
5 Greenleaf Woods Drive, Suite 102
Portsmouth, NH 03801

Book design by Tim Holtz
Printed in the United States of America

For my Dad

"Jack" Goethel, an American original, circa 1988. Photo by Ellen Goethel.

Contents

CHAPTER 1

"Stay With Me"

You never die the way you expect to. This was what I assumed would be my last conscious thought as I hurtled from the tailgate of my truck toward the pier and boat some twenty feet below. I am, or was, a commercial fisherman unloading my boat using a hoist at the Hampton State Pier in Hampton, New Hampshire, on a bright sunny, early fall afternoon on September 15, 2010. My boat is a forty-four-foot Stanley stern trawler custom built for me in the winter of 1982. My name is David Goethel, and I am the owner-captain of this vessel. We were participating in the silver hake (locally called whiting) fishery and had completed unloading our whiting on the other side of the harbor at Yankee Fishermen's Cooperative some minutes before. We were now unloading fifteen one-hundred-pound boxes of mixed sea herring and red hake to be trucked to Portsmouth, New Hampshire, to be sold for lobster bait.

This was a routine occurrence, performed thousands of times over the years. So, what went wrong? Well, to explain, the reader needs to understand the layout of the pier and hoist. The pier is a steel bulkhead made of steel sheet piling driven into the mud at the edge of the harbor. There are floats, which ride on pilings against the sheet piling. The floats are four feet wide and have a narrow ramp at one end for people to access the parking lot that abuts the sheet piling. There are ladders roughly every hundred feet, but they are wood with the rungs held in place by rusty nails, which no one trusts. A fisherman had been hurt several years ago when, halfway up the ladder, he grabbed a rung that pulled out, sending him crashing to the float below. In fact, the whole system was run-down, having been built in the mid-1970s and having received no serious maintenance since. On the top edge of the sheet piling was a twelve-by-twelve-foot wooden beam bolted to the steel. This likewise was in poor shape and wobbled back and forth if someone stood on it. The hoist is a fifteen-foot vertical I beam cemented into the parking lot. Approximately ten feet above the ground is a horizontal I beam approximately twelve feet long. There is a trolley mounted on the lower edge of the beam to which is attached an electric chain hoist capable of lifting one thousand pounds. The hoist is powered by a 220-volt electric motor whose wires run by conduit up the I beam. The up and down control for the hoist is at the end of a fifteen-foot cord allowing the user to stand in the nose of his truck or well away from the edge of the pier. The horizontal beam is mounted on a swivel pin so the arm can either swing out to pick things off the boats or swing towards the parking lot to place items in vehicles. The fishermen had tied a length of rope to the end of the horizontal arm so they could pull the arm to a desired position. Unfortunately, someone, maybe a fisherman, a recreational boater, or a teenager using the hoist as swimming device had pulled on the cord, which broke the wires to the control. This happened frequently and the state had to cut the wire back and remount the ends. This time the person cut about nine feet off the wire, shortening it to a dangerous length. Numerous fishermen had told representatives of the state that this was an extremely dangerous situation, and someone would get hurt. I drew the short straw.

September 15 was right at the time of full moon. On full moons, the grav-itational pull exerted by the moon is at its strongest. Thus, the high tides are extremely high and the low tides extremely low. The tide was two feet lower than normal, making the distance from the tailgate of my truck approximately twenty feet instead of the usual fifteen. Either way, it is a long way to fall. We had successfully unloaded two lifts of three boxes each when the hoist swung out and the control dangled out over the harbor. I went to reach for the control with my right arm and missed it. The momentum of my arm coming down caused me to lose my balance and begin to fall. In a last-ditch effort to save myself, I tried to land my feet on the twelve-by-twelve. It pivoted out toward the harbor and my fate was sealed. Applying simple gravitational physics, it took me a little over a second to reach the float. That does not seem very long, but your mind races and time seems to slow down. I had seen recreational boaters' watermelons role off the edge of the pier to the float below. Near low tide, the results were always the same, an explosion of red and green. I assumed the same would happen to my head. While falling, I thought about my wife and kids and this unexpected turn of fate. Then BAM! Everything went black. What follows is what I was told by several different people involved in the aftermath of the fall. I remember nothing for a period to this day.

My crewman was named Jeff Emerson. He had worked at the Yankee Coop unloading fish while going to UNH to study engineering. He had never fished but took a job on the *Ellen Diane* when the last crew disappeared. Jeff had no experience dragging and at first got very seasick. However, he was tough and determined. Through proper medication and plain will power, he overcame the sickness and learned the job. Jeff had become adept at the job, getting the net safely in and out of the boat, sorting, boxing, cleaning and icing the fish and securing the boat and unloading in the harbor. I liked Jeff, and even though I knew he had no intention of staying if he got a job in his field, I was glad to have him as crew.

His recollection of events was that I pivoted off the pier hurtling head down towards the float. He thought I tucked my head just before impact and my left shoulder and head hit the float simultaneously. Jeff fumbled for his phone to call 911 and summon help. Then he dialed my home phone and told my wife, Ellen, there was a terrible accident and she should come to the pier immediately. I lay on the pier unconscious, blood pouring from every orifice. The fire station is on Hampton Beach, so he could already hear the sirens when suddenly I stood up and muttered something that sounded like "that did not hurt as much as I expected," took several steps, and passed out again. Now Jeff was truly scared. The boat was tied securely against the pier, which meant I could not roll off into the water. However, I had almost staggered to the stern of the boat. We had practiced what to do in a man-overboard situation but had always assumed the person being rescued could aid in that rescue. Should I enter the harbor unconscious with God knows what kind of injuries, he would be forced to jump in and try to hold my head above water until help arrived. Just when he figured things could not get worse, I staggered to my feet again and went about ten feet past the end of the boat before collapsing. This time I stayed out for about twenty minutes.

When I first came to, I assumed I was dead. I was trying to ascertain whether I had taken the up or down escalator. My first thought was I must be in hell. It smelled like sulfur, i.e., brimstone, and I could hear a boat running. I was quite sure there would be boats in hell because the damn things never ran right, and you could spend eternity trying to fix them. The pain was unbearable, and it sure seemed like eternity was going to be a long time. Still . . . it could be heaven because lots of people liked boats, and there was nothing finer than a beautiful summer day on the ocean. However, Saint Peter had some serious work to do cleaning up the entrance to the pearly gates, and I did not see anyone with wings. So . . . I must still have been on earth, which meant I was alive! I needed further confirmation, so I tried to perform a self-assessment. I could see, what could best be described as looking down a pencil, out of my right eye. I could not move my head, so my field of vision was part of my right arm, an arm with white, peach fuzz hair trying to keep me from moving and the port corner of the stern of the *Ellen Diane*. I tried to talk to the peach fuzz, but nothing came out. I could ascertain from his voice he was nervous, and I felt like I was in some TV drama where the people keep saying "stay with me." The pain was immeasurable, and I had to use every bit of willpower to fight through it. Every time I tried to sleep, I heard, "stay with me." My next thought was to see if I was paralyzed. I could move my right arm and wiggle my right toes. I guess they had already removed my shoes. I could not move my left arm and through great pain could barely move my left toes. All right, so I wasn't paralyzed, yet. I was breathing on my own, but my throat was full of blood, and I felt like I was drowning. The left eye did not work, my left ear was either missing or severely damaged, and my teeth were mashed to the right side of my face. Still, there appeared to be a slim chance, if nothing else went wrong, that I might live.

The paramedics and firemen now faced a paradox. The ramp was too narrow for the gurney and there was no way to carry it at the steep angle produced by the tide over the railing. I could hear muttering and debate about how to proceed. Basically, it came down to fashioning a sling, putting me on the gurney, and trying to lift it with the hoist, or using a backboard, which looks like a surfboard with handles cut in the edges. Ultimately the backboard won out. The next problem was getting me on it without bending my back, neck or head, which was immobilized in some device that kept it from moving. The assumption was I had a broken back or neck and had a crushed skull. Some time had passed, and the cobwebs had cleared a little. The men told me how they were going to move me and that there would be lots of pain. They were not kidding, but with one muffled scream I was on the board. Now they still had to get the board up the ramp, which had a roughly forty-five-degree angle, while keeping me straight. They carried me with six men the way pallbearers carry a coffin. At the ramp, the

End of the day. F/V *Ellen Diane* entering Hampton Harbor with a load of fish. Dan Goethel on the back deck, 2011.

two middle men dropped out, and somehow the remaining four lifted my two-hundred-pound frame and kept me level for placement on the gurney near the ambulance in the parking lot. As I was placed on the gurney, I heard a voice from the crowd, "David, I am here for you, I will always be here." The voice sounded calm and cool under pressure. The voice belonged to my lovely wife, Ellen.

The paramedics and firemen of Hampton deserve a great deal of credit. I have no doubt I am here today because of their extraordinary professionalism under extreme conditions. Now began the odyssey of recovery. I had never ridden in an ambulance and had always assumed they would just glide down the highway like an oversized Cadillac. Nothing could be further from the truth. They bang and rattle and every crack in the road feels like you hit a pothole. Exeter Hospital in Exeter, New Hampshire, is about seven miles away and when we arrived the medical drama continued. Just like in the movies, when we arrived, medical staff surrounded the gurney and began figuring out how bad the damage was. Clothes were cut off and we rushed down a hall into a room with some giant machine. I was fed into the machine, and it whirred and clicked, and I was removed. I later learned it was a CT machine. I think I was also X-rayed, but I am not sure. Then we waited a while. Whatever these assembled machines showed, apparently the picture was not good. The doors flew open, and we moved rapidly down the hall we had entered from, with medical professionals telling people to clear out of the way. At the end of the hall, my one working ear picked up the low thump, thump, thump of helicopter rotors. Shit, I would get a chance to fall out of the sky twice in one day!

I was slid through the side door, strapped down, and off we went. My grim sense of humor kicked in, as I thought, at least I do not have to ride in a plexiglass container on the runner like the wounded in *M.A.S.H.* I do not know much about medicine, but I know if you're getting helicoptered to another facility, it is either in Portland, Maine; Dartmouth, New Hampshire; or Boston, Massachusetts. By the angle of the sun, I determined we were headed for Boston. A man with a plexiglass helmet that made him look like an astronaut rode beside me and only spoke to tell me not to move my head. The sun was setting when we landed on the roof of a hospital in Boston. Just like in *M.A.S.H.*, the doctors were in the helicopter before it even touched the roof. We went through doors, into an elevator, and down for more tests, many more tests. Meanwhile needles and tubes were being stuck in places I did not know you could put needles and tubes. Doctors and nurses asked me questions, hundreds of questions, and then ordered more tests based on the answers. As fishermen, we leave the dock at 4:30 a.m. every morning. It was well after dark, and I was tired and thirsty. Nobody seemed to care about the pain. I learned later that was because of the head injuries. I had bleeding on the brain and over one hundred cracks in my head as well as two shadows on my neck vertebra, which at the time were considered possible cracks. In short, this was serious. If the brain bleed did not stop, the neurologist would have to drill holes in my skull to relieve the pressure.

Meanwhile, my wife had been left behind in Exeter. She had contacted her sister for a ride to Boston, figuring rightfully, she was in no shape to drive. On the road, she contacted my oldest son, Eric, who was captain of a tugboat in Boston Harbor and filled him in on what had happened and where I was. Contacting my other son, Daniel, was more difficult as he was boarding a plane to travel to France for a scientific conference. The Massachusetts state police located him on the plane, and in what must have terrified the other passengers, escorted him and his companion off the plane. What they did next, I still find extraordinary. With sirens blaring, they rushed him from East Boston, through the notorious Boston rush-hour traffic, through the tunnel, and through downtown Boston to Beth Israel Deaconess Hospital located west of Fenway Park near the Brookline city line. Knowing Daniel would arrive first, my wife planned with the hospital to have a social worker meet him and provide details and comfort.

Ellen and her sister, Sue, arrived next. They were met by Daniel and his then girlfriend, Annie. Sometime later, Eric arrived and then Sue's husband, Charlie. I tried to break the tension by telling them the only thing left that was not broken was my sense of humor and sarcasm, and I would be damned if anything else was getting broken today. Gradually the extent of the devastation came into focus. Both of my left leg bones were cleanly fractured above the ankle. My left scapula, which is the large bone in the back of your shoulder, was broken. This could not be set. Instead, my left arm was immobilized, and I was propped up in such a way that there was little force on it. My entire left side was badly bruised, but there was no internal organ damage. My cranial nerve, which comes over the top of your head and splits—half going to the left side of the face and half going to the right—was damaged but not severed. I had no control of the left side of my face and could not blink my left eye. This was serious because blinking keeps the eye lubricated and, without that capability, there was very good chance I would go blind in my left eye. No one knew it yet because of the device that kept my head from moving, but my left ear was only attached to my head by a small piece of flesh, and my inner ear bones were scrambled. The main issue was the bleeding on the brain. That had to be controlled or everything else was irrelevant.

Doctors who were specialists in specific body parts rotated through in groups of six or more, as this was a teaching hospital. They stood to my left and talked quietly about me. Nothing on my left side worked and I grew frustrated with all these people talking about me but not to me. Finally, some poor intern came over on my right side. My right arm was my only functioning body part. I grabbed his necktie and pulled his face into my very small field of vision. I explained succinctly that everything to the left of center was broken and that I wanted them to come on my right side and talk to me not about me. The poor man had the deer-in-the-headlights look, but after I released him, everyone came to my right side.

As the evening wore on, my assembled family became hungry, at least the men. Somehow, they got pizza delivered. It smelled heavenly, but I was not allowed to eat. Finally, the nurses politely, but firmly suggested that my family leave for the evening. My wife protested mightily, but to no avail. When she returned early the next morning, I was having a seizure. From then on, she slept in a chair at the end of the bed until I was discharged.

Bingo, Bingo, Bingo!

The date had probably changed in the Intensive Care Unit, but I had no way of knowing. What I did know is the pain emanating from virtually every part of my body was increasing exponentially. Two areas were particularly annoying: my left leg and foot—which had been placed in what looked like an oversized ski boot that stuck out the end of the bed and was hit by every person as they walked by—and the area under the brace on my head in the vicinity of my left ear. I was really beginning to feel like I might lose my mind, but everyone I asked to look under the brace said the brace could not be moved. Frustrated, I finally took matters into my own hands. I worked my right hand loose from the bedding even though it had several tubes stuck in the veins and gradually got enough slack tubing in each portal to get my hand to the brace, which I attempted to remove. A nurse caught me and pushed some button to summon reinforcements. A doctor told me sternly that I might die if the brace was removed. I responded, fine, I would rather be dead then endure this pain any longer. He sighed and very carefully tried to lift the corner nearest my ear. In classic doctor speak, he mumbled an "un-hah" and said he needed to make a phone call. A few minutes later a whole gaggle of new people arrived led by a plastic surgeon, and the neurologist reappeared. The brace was very carefully removed, and I was given very clear instructions not to move my head. At this point there were several "ah-ha's" and one "oh my." With all my other injuries, no one had noticed that my ear was only attached to my head by the very lower part of the ear lobe. There was plenty of both fresh and dried blood. A decision was made to reattach the ear and to use a topical anesthetic lidocaine. A young lady resident was selected from the crowd to do the attachment. A discussion ensued between the resident and the plastic surgeon as to how many stitches to the millimeter should be performed to avoid scarring. Meanwhile having removed the pressure from the brace on the wound, the pain jumped to new levels of torment. My little bit of remaining restraint finally gave out, and I told the teacher and student as politely as circumstances warranted that I was fifty-seven years old and ugly when I arrived, and I expected to leave fifty-seven years old and still ugly. Now, get on with it! The ear turns out to be a particularly tough piece of cartilage and the poor young lady could not force the needle through it. The lidocaine had no discernable effect on the area whatsoever and I screamed. The plastic surgeon finally took over and completed attaching my ear. The internal damage would be dealt with later. The brace was reapplied, and the pain subsided slightly.

Sleep came in very short spurts, if at all, and as light began to seep through the window all the doctors started their rounds. So far, I'd had teams of neurologists, orthopedic surgeons, ophthalmologists, and plastic surgeons. A team of ear, nose, and throat specialists would be added later that day and my dentist was contacted. A pretty impressive collection of people were working to put Humpty Dumpty back together again. Nothing really changed during day two, but by day three the neurologist was glad to report that somehow, my brain bleed had stopped, and my body was resorbing the blood.

More tests were done to confirm this information, and the next day I was moved out of the ICU into a room to begin my recovery. Troops of doctors came through several times each day to check on my progress. One group I found humorous were led by the head doctor of the floor who bore an uncanny resemblance to Marcus Welby. He led a group of students whom he grilled thoroughly about various aspects of people's cases. When a student answered to his satisfaction, he would boom out loud enough for the whole floor to hear, "Bingo! Bingo! Bingo!" Many of the students struggled to keep their composure, and I am sure they wondered what the patients must think. I started to say once, "I will take brain bleeds for 200," but thought better of it as it might not be received the right way.

Several doctors and psychologists from the neurology department spent many hours trying to learn what areas of my brain and nervous system had been impacted. For example, they already knew the nerve that controlled muscle movement in the left side of my face was not functioning, but the nerve that controlled pain worked just fine. Just my luck! From these tests, I also learned that my short-term memory for names and numbers was severely impacted, but my long-term memory appeared to be very strong. They asked me to try to recall specific times in my life, beginning as far back as I could remember. All the questions were designed to test various aspects of brain function. When I was not being quizzed, I was instructed to try to remember my life. This served two purposes: to keep my brain active and functioning, and to keep my mind focused on something other than the constant pain, which could rob people of their will to live. I had already figured out this second aspect and had begun intently staring at specific body parts through my one good eye and willing them not to hurt for five minutes. The part initially had to be small, say a toe, but over the ensuing months I worked it up to entire appendages. Five minutes, however, remained about the maximum time.

What is presented in the following chapters are those recollections, for which I wrote out outlines as I regained the use of my left arm in the following months.

CHAPTER 2

Early Years

I could be considered a poster child for the baby boom generation and experienced all the joys and tribulations of what was then the United States' largest cohort of people. To recite some genealogy, I was born in 1953 to Barbara and John Goethel in Boston, Massachusetts. My father was a survivor of World War II, having been shot down over the North Sea in a fall gale. My mother had worked for the Red Cross during the war, and they met when my father was discharged from the service in Boston. They shared a cab back to their parent's homes in Newton, Massachusetts, and the rest as they say, is history. They shortly married, settled in Needham, Massachusetts, and began raising a family.

My father, like many men from the war, came home with what would today be called post-traumatic stress disorder. He self-medicated with whiskey and cigarettes. He earned a degree in law on the GI Bill and worked for the John Hancock Company. My brother, Fred, arrived in 1955.

In 1962, at the age of nine, an event occurred, which, at the time, did not seem particularly noteworthy, but which ended up determining the direction of my entire life. My father took my brother and me deep-sea fishing in Seabrook, New Hampshire. I do not know how he located this harbor, which was more than sixty miles from our house. After all, there was no internet, no Google, and I doubt they could afford advertisements in the Boston paper. At any rate, we set out early one morning and located two businesses on River Street in Seabrook, New Hampshire. I asked my father how he was going to pick who to go with. His response was we would go with the ones with a garish blue-and-orange sign. My father loved eating at Howard Johnson's restaurants, which the restaurateur had made famous by painting them the instantly recognizable ice blue and bright orange. My father figured given the restaurant's success, these people must at least have marketing savvy and would be worth trying out. It was the right decision. We went out on Bill Eastman's boat, and I caught nine mackerel. I had a new occupation to write about: when I grew up, I wanted to be a fishermen.

"I got dinner!" David Goethel *(left)*, age ten, holding a pollock. His brother, Fred *(right)*, in Gloucester Harbor, 1964.

Being in the hospital in Boston, inevitably led me to remember another traumatic medical episode. During the summer between third and fourth grade, I somehow contacted viral spinal meningitis. Numerous doctors were consulted, and since the disease is rare, no firm diagnosis could be made. As my fever surpassed 105° F, I was brought to Boston Children's hospital. Children's is one of the finest research hospitals in the world, but because they do research on rare diseases they needed samples, lots of samples. The assembled doctors decided a spinal tap was necessary to confirm their suspicions. Most meningitis is bacterial. This has a high fatality rate but is treatable with antibiotics. I had an unrelenting headache, and the fever was still climbing past 105.5. If you have never had a spinal tap, I cannot convey the intensity of the pain. A huge number of doctors and nurses assembled to hold me down and immobile on a cold metal table. A huge needle was inserted into my spinal column in my lower back to extract fluid. I screamed like I have never screamed in my life. My parents left the room in tears after the first needle. The needle was inserted again and again. I finally passed out. I do not know how many samples were taken and how many were necessary for the diagnosis, but I have always believed that more samples were taken than necessary for research. I woke up in a room surrounded by people in moon suits. Even my parents had to wear them. I was told later that the doctors had bluntly told my parents that I was expected to die, and that, because they did not know how the disease was transmitted, all necessary precautions must be taken to avoid the disease making it out of the hospital. Various treatments were tried, such as submerging me in a bathtub of ice and water to break the fever, which peaked at 105.7, along with pills and potions, but nothing worked except time. Another child was brought in and shared the room behind a curtain. I later learned he had incurable cancer and died after three days. He was placed in with me because a cold calculus was made that he would die from the cancer before he would contact the meningitis. Gradually the fever dropped, and when it got into the 103–104 range, I began to hallucinate. There were cracks in the ceiling that kind of looked like a dinosaur, and when it was dark, I believed it would eat me. I begged the nurses not to turn out the lights, and they obliged me, although they had no idea why. After several weeks, the fever and headaches disappeared, and I faced my next challenge. I was so weak, I could not walk and needed physical therapy to relearn how. About four days later, I could move to a wheelchair, and the doctors determined my convalesce could continue at home. This brought the next great test of wills. My parents had brought my stuffed animals to keep me company. The doctors said everything in the room must be burned, including my clothes and stuffed animals. Sensing a test of wills between a stubborn, sick child and the medical establishment, my father asked everyone to leave the room so he could talk to me. He told me that he would spirit out my two favorite stuffed animals in the plastic bag he had brought the clothes I was to wear home in, and the rest would go to the incinerator. Apparently, you cannot get viral meningitis from stuffed animals as those two were in my bed for several more years.

Most of the summer was over. I worked on walking every day, but I was listless. I did not want to play with my dog, Gus. My parents were worried. Someone came up with the bright idea of buying six chicks and keeping them in the old farm shed about one hundred feet from the house. No one said it at the time, but the rationale was I would have to walk back and forth to the shed to feed, water, and otherwise take care of the chicks. Now I had a cause, and it helped immeasurably. My walking improved and my flock grew. At its height, it would number some thirty chickens, half a dozen ducks, and twenty pigeons of various varieties. Maybe I would be a farmer.

Fourth grade passed uneventfully, but I would remain weak and underweight through eighth grade. My dad and I went fishing more and more at Eastman's and off docks up and down the coast. The world was changing, I just could not sense in which way. As a child, you sense when your parents have problems.

Fifth and sixth grade came and went. I was an above-average student but not exceptional. I did well in subjects I liked such as history and science and had excellent reading skills but poor spelling and low math skills. My dad took me fishing every weekend for the summer, and we were now regular customers on the stern of the party boat. The regulars all knew each other and brought food and alcohol to be consumed while fishing. They were good fishermen, including several wives, and a friendly group of people. Later that year, I graduated elementary school and prepared to go to a newly constructed, enormous junior high school. My only solace was in fishing. I paid very careful attention to all the boat activities and began to immerse myself in what would become a career.

Junior high was new in many ways. The school was only two years old, but it was huge, holding over eighteen hundred students. The people of Needham were obsessed with sports. Both boys and girls were expected to participate in something. The only people who got a pass from that expectation were those students who participated in band. Anyone else was soon labeled as socially deficient by the lords of junior high. The first step for boys was the administration of a physical fitness test that scored you based on your height and weight. Remember now that I was a year younger than everybody because I started school a year early, and I was still severely weakened from meningitis. The test was administered in front of your gym class and had things like pull-ups, push-ups, and some spring-loaded device with a handle you pulled up on to measure upper and lower body strength. A combined score of one hundred was average and most boys scored between ninety-five and one hundred and thirty-five. My arms and legs were still severely atrophied, and I could do neither push-ups or pull-ups and the spring machine moved the needle about half of what most people did. I scored a sixty-four, which spread around the school slightly slower than the speed of light. Thus began the torment of junior high. I picked up a pair of bullies, one who was short but muscular and on the wrestling team and the other tall and gangly and apparently eager to prove his manliness so he would not suffer the same fate as me. Seventh grade was not bad, just annoying, but by eighth grade it had morphed into serious physical threats.

The summer between seventh and eighth grade, my dad and I fished every weekend and most of his four-week vacation. My brother went occasionally, but he really had no interest in fishing. My mother was spiraling into crisis, seldom left the house, and never left the yard. On the party boat, I was given the title of second mate, which meant I helped grind chum, a fish called sand eels put through a meat grinder and used to attract the mackerel—untangled lines, and helped keep the boat clean. Other than being a boost to my ego, the chief benefit was my father did not have to buy a ticket for me. I continued to learn with the goal of graduating to first mate when I was old enough to work, which was a paid position. There were two mackerel trips per day, 8–12 and 1–5, but you could not fish for mackerel through the entire season because they migrated south when the water temperature dropped below 48 degrees. During those periods, the Eastmans ran an all-day boat that traveled much further offshore to fish for cod, haddock, pollack, and other bottom-dwelling species. This was an entirely different type of fishing, using deep sea reels with heavy line and one-pound weights to take your line some two hundred feet to the bottom. The bait of choice was sea clams cut into bite-sized pieces. Another difference with these trips was that the crew ran a pool in which anyone who contributed a dollar could have his fish entered. The heaviest fish won the pot of money. The winner was expected to tip the crew for collecting the money and weighing the fish. I quickly learned two things. First, the deep-sea reels were made for right-handed people and the handle was on the wrong side for left-handed people. Fishing and having to crank with the wrong hand were challenging. Second, the people who usually won the pool used artificial lures called jigs that they purchased. The jigs simulated a wounded herring and needed to be continuously bounced up and down with the rod. The cheaper jigs were made of lead and painted with chrome, but there were also stainless-steel jigs of various sizes, imported from Norway, that were particularly effective. My dad and I never won a pool that year, but we learned a lot and became properly equipped through the following winter.

I started eighth grade in 1967, which proved to be one of the most tumultuous years of my life. The bullying started immediately and increased both in intensity and ferocity. I was not the only person having issues. Most days there was a fistfight either in one of the halls or the cafeteria. Some of these were truly savage and the male teachers had a hard time separating the combatants. Junior high seemed to be a microcosm of the whole country as society seemed to be coming unglued.

In the middle of winter, I finally snapped at my tormentors in a school hall. They knocked my books out of my arms, and I turned and screamed at them using words we could charitably describe as undiplomatic. This was observed by a teacher who marched the three of us to the assistant principal's office. The teacher reported his version of the events, parents were called, and everyone got detention. Mine was for offensive language, which "shocked" my mother. The other two were for fighting. I assume the two bullies got a severe dressing down from their parents or worse, because the first day their detention was over, I was met on the walk home through the snow by the two amigos, who tossed my books as far as they could into the snow and then administered a beating. I tried to defend myself, but I really had no idea how to fight and finally arrived home bruised and battered. (It was strange how the memories of injuries past kept popping into my head as I lay in the hospital recovery room. Maybe it had an "I've had worse" sort of effect on my being able to deal with the new injuries?)

I asked my father to teach me how to fight, but was informed sternly by both parents to turn the other cheek. I vowed at that moment to ask Captain Bill next summer to teach me what to do, as I had seen him restore order when recalcitrant passengers got out of control. I am sure my parents meant well, and my father, as a lawyer and representative of the court, could not condone illegal behavior. However, just as a judge is never present when an abuser walks through a restraining order, the law was not going to fix this problem.

With the onset of spring, weekend fishing resumed, and I told Captain Bill about my problem. Bill had been an amateur fighter in the navy in World War II and clearly knew how to box. The basics were explained: how to make the opponent think you were going to lead with one hand to draw him off guard so you could connect with the other. He told me I would have the advantage because I am left-handed and most people expect you to lead with your right hand. We practiced and he built up my confidence with words of encouragement as well as an admonition not to start fights and not to fight unless there is no other option and you have been severely provoked. The lessons occurred the first three weekends in May. School was still in session and my tormenters must have decided to have one last laugh at my expense. Once again, they jumped out of the bushes and tried to grab my books, only this time things were different. I threw down my books and sized up the shorter one because he was the meaner of the two. I faked with my right hand, he brought his left arm up to block leaving his nose and entire right side of his face exposed. I unleashed my left fist with all the pent-up rage of two years of torment and got him squarely in the nose and right eye. Blood spurted from his nose and my left hand hurt like hell. The other kid looked stunned and was probably considering his options, but there were none. I turned to him, pulled the same move and connected with the same result. They both ran off screaming and one was crying. I saw them in school the next day, both sporting bandaged noses. I always wondered what they told their parents. I never had a problem with them again, or anyone else for that matter, in junior high or high school.

Summer passed quickly, and I gained more and more experience on Bill's boat. All of their boat names began with the word "Lady" and they called themselves the lucky lady fleet. Bill's boat was named after his wife Lil. They had met and married at a very early age. Lil was thirteen and Bill was fifteen. They had a son Bill Jr. a year later. Broke and in the Depression, the young family drove themselves cross-country to California where they fished for tuna as a couple. World War II started, and Bill enlisted in the navy where he fought in the Pacific theater. At the end of the war, the young family moved back to Seabrook, where they started the party boat business with Bill's father, Oscar, and mother, Mae. Oscar and Mae had a second son, Lester, born during the war. Oscar suffered a stroke in the mid-1960s, leaving his half of the business to Bill and Lester. Lester ran the all-day boat named after his wife, *Lady Marcia Ann*. Lester and Marcia likewise married young and had three sons and a daughter spread over a ten-year period in the 1960s. Bill's son Bill Jr. ran a boat named after his grandmother Mae. He graduated college and with his wife, Anne, had two children. Bill and Anne both taught school in the off season. The remaining two boats were run by hired captains. Max ran the *Lady Rosa*, named after Bill's sister, and Dick Muise ran the *Lady Bev*, named after Bill's daughter.

Late in the fall, just before the end of the season, Dick Muise's crew failed to show up. I was offered the chance to be first mate on an all-day charter. My dad would go along as the official mate because I was still too young to legally work. Dick had an interesting back story. He was a native of Prince Edward Island, Canada, where he had been a commercial fisherman in a small village. A hurricane struck the area while the small community fleet was at sea. Dick's boat was the only one to return. Rather than rejoice that anyone had survived, the village turned on him thinking he must be in cahoots with the devil. He was forced to flee the island and settled in the United States on Plum Island, Massachusetts, where he ran various vessels.

He was in his first year working for the Eastmans. The fishing was typical for early October, which is to say, not that good. Dick asked me to fish, both to show people there were more fish than they were catching, and to give them the fish I caught to take home. I always wondered how many men returned from these charters, slapped a big cod on the table, and told the wife, "A thirteen-year-old kid caught it and gave it to me." I assume that number would be zero, but it was good public relations. I caught a few nice haddock and a couple cod, but since I did not get offshore that often, elected to try my new stainless cod jig. I put a white worm above it. This is a long-shanked hook run through a piece of surgical plastic tubing. I chose white because pollack seem attracted to the color. Pollack are a close relative of the cod. They move to the west in the Gulf of Maine to spawn in November and December. Pollack are a very powerful fish, offering much more fight than cod or haddock. Many anglers have reported they thought they had hooked a large halibut when they had a double on big pollack, one on the worm and one on the jig. I lowered the rig to the bottom and jigged for about ten minutes with no success. Pollack are semi-pelagic, that is, they do not spend all their time on the bottom. I reeled the jig about fifty feet and let it fall back to the bottom. This produces a different motion for the jig and the worm then just bouncing the jig up and down. About halfway down the jig stopped sinking. What was going on? Had my knot come untied? Had I lost the jig? I flipped the free spool lever back on the reel and before I could start cranking the rod was nearly ripped out of my hands. The fish was taking line steadily off the reel and heading off at an angle. Clearly no cod, which have about as much fight as a boot full of water. Either a halibut or double on pollack. Either way, it was the epic struggle every angler dreams of. I soon had everyone's attention. Most of the charter reeled in their lines out of respect for the battle unfolding and not wanting to be responsible for tangling the line and losing the fish.

The fish now had about one hundred feet more line dragged out then when the battle started. If it did not turn soon, I would have to tighten the drag further or run out of line. Tightening the drag risked snapping the line. Finally, the fish turned and I gained some line. But it was not ready to capitulate. The fish and the angler went back and forth for about twenty minutes before the angler began to gain the upper hand. The fish's circles became shorter; the fish was tiring, but so were the angler's arms. Back and forth we went until a faint silver glow began to appear deep in the water. The glow was fish-shaped, not flat like a halibut. This was a pollack, but why was it so epically strong? The answer soon became apparent. Dick leaned over the rail and sunk a gaff in the fish's head. Then another unusual event. Dick struggled a little bit. He was a big man, over 6 foot 2, and he was hanging onto the gaff with both hands as the fish thrashed him around. When Dick finally lifted the fish over the rail, the reason for the epic battle became apparent. The fish was huge, almost as long as I was tall. We weighed it on the boat at thirty-eight pounds. Pollack do not get as large as cod and upon reaching shore we learned the International Game Fish Association (IGFA) record was then thirty-three pounds. I had a world's record! When we reached shore, we had the fish weighed on a certified scale at a lobster pound. The official weight was thirty-seven pounds, three ounces. Every measurement, length, girth was taken along with numerous pictures. The rig was stored away to be produced if necessary, and my dad would send away to the IFGA for whatever forms were necessary. The fish had both the jig and the worm in its mouth. Whether it struck the worm first, and because the line was slack, swam down and bit the jig as well, or it just got tangled up in the fight no one could say, and it would not matter anyway. The fish did not qualify for the record because you can only use a single hook. This was still quite a day!

Since the pollack was caught on a Sunday, my dad and I had to drive back to Needham so I could attend school on Monday. All this extra activity at the pier had chewed up substantial time and we did not leave Seabrook until about 8:00 p.m. The ride down Interstate 95 and Route 128 usually took a little over an hour, but this was foliage season and the traffic moved slowly. We did not arrive in Needham until about ten. My mother came from a family of educators, and, to her, school was more important than anything. On school nights, we were supposed to be in bed by 9:00 p.m. Not only did she not want to hear my fish tale, but she also started a heated argument with my father about being late. It is hard to believe that one fish could dissolve a marriage, but the pollack was emblematic of much bigger problems in their marriage and the marriage essentially ended that night. It was a lesson I took to heart when choosing my own life partner in the years to come. To compound matters, I was in ninth grade and had been invited to take advanced English. There was a lot of reading and twenty new vocabulary words on which you were tested every day. I was always a lousy speller and struggled to get C's on these tests. Monday, I took the vocabulary quiz on words I had not studied and scored an F-. My mom hit the roof and another huge argument ensued. By the following February, the marriage was over.

A family meeting was called in late February. My mother was crying, and my father's jaw was set. He announced he was moving out into an apartment for a trial separation of six weeks. He never slept in the house again. The rest of the year was grim. My mother did not drive, and even when her friends offered to take her shopping, she refused. There were no stores within several miles, and I was not old enough to drive. My father charitably brought groceries on the weekend, and my mother figured how to order whiskey and cigarettes by the case and have them delivered by a liquor store. My brother and I were stranded there for the week. The situation did not bode well for the future. In March, I received a call offering me a job with Dick Muise on the *Lady Bev* as first mate. There were two requirements: I would need to find a place to stay and produce what are called working papers in New Hampshire to satisfy the labor laws.

My father stepped in and both drove me to New Hampshire to procure the working papers and made a monetary arrangement with Marcia and Lester Eastman to feed me and allow me to sleep in their house for the ten-week summer season. He would drive me back and forth to Needham during school time in spring and fall. Marcia and Lester were nice people, and I enjoyed hanging out with their two oldest children, who were only slightly younger than me. Dinner was just another plate at the table, but their house was small, and I always felt awkward taking up sleeping space.

Massachusetts had instituted "no fault" divorce that year and my parents soon took advantage of it. My mother got custody of us, alimony, and child support until we were both eighteen. I graduated from junior high school, packed a suitcase full of clothes, and moved to New Hampshire.

Being a mate entailed getting the customers rigged up to fish, untangling their lines, removing spiny dogfish from their hooks so they did not get hurt, hauling the anchor when the captain moved the boat, cleaning the boat at the end of the trip, and making sure provisions such as hooks, sinkers, toilet paper and the like were stocked in sufficient quantity. I also got to steer the boat a lot so the captain could rest, and I became quite proficient at sticking to the compass heading selected to make sure we arrived at our destination. I earned ten dollars per day, which was held in the company safe until summer was over. By Labor Day, I had amassed over five hundred dollars, which I spent on a high-quality reel-to-reel tape recorder to record music I enjoyed off the radio. I still have the tapes and the recorder. I also learned that even though you earned the money, you do not keep it all. I paid my first income tax and well remember my distaste at the idea that someone else got to spend the money I'd earned.

Marine travel lift launching the Lady Lil II in 1976 Newburyport MA.

CHAPTER 3

Death and Taxes

I became schooled in taxes and learned a little about death during two events on the ocean that year. The first occurred on a late July day at the Isles of Shoals, a set of small, rocky islands twelve miles from Hampton Harbor. As I was learning, when the water warmed up near shore, the mackerel tended to congregate in places where currents caused colder, deeper water to be moved to the surface. This occurred regularly around these islands where the tides ran swiftly. Consequently, the mackerel boats from Rye, Hampton, and Seabrook fished here almost daily in the late summer.

This day was like so many others as the *Lady Bev* pulled into anchor under the lighthouse on White Island, the closest island to Seabrook. The anchor was set, and the people began fishing. Other party boats were anchored on both sides of us, and others were anchored off the faces of other islands. Captain Dick eyed the sky nervously, as it was darkening quickly to the west southwest. The wind was moderate out of the southwest and the stern of the *Lady Bev* was less than several hundred feet from the granite shore. Even to a novice, it was apparent we were in for a thunderstorm. Party boats are required to carry ship-to-shore radios, which are always tuned to the international call and distress frequency, 2182 khz. The weather service issues special marine warnings through the Coast Guard on this channel. None were given this day; if they had been, we would have hauled anchor and moved away from the lee shore of the island to have sea room. About fifteen minutes later, the wind picked up dramatically and rain began. You could see lightning and hear thunder in the distance.

Captain Dick told me to remove the scupper balls from the scuppers and told the people to keep calm and shelter under the topside canopy that extended from the pilothouse about twenty feet towards the stern from rail to rail. I was further instructed to keep the scuppers clear from the torrential rain that Dick expected. "Scupper" is a nautical term for holes cut from the edge of the deck to the outside of the hull to let water from a boarding sea or rain flow back to the ocean. They are kept plugged in nice weather so that water does not shoot from the outside to the inside, soaking the customers feet. Dick let more anchor line out, which again, nautically, is called "scope," to make sure the anchor did not drag in the increasing wind. He informed me the back deck was my responsibility, and he would start the engine in case the anchor broke loose and handle the anchor.

I watched a two-masted sailboat frantically trying to drop all its sails in the channel that passed between White and Star Islands. Then all hell broke loose. The wind picked up to levels I had never experienced and then picked up some more. Hail pelted the boat along with torrential rain. The ocean became chaotic with spray blowing in from all directions seemingly at once. Lightning hit the lighthouse several times, the flash momentarily illuminating the chaotic scene. The noise was deafening, and you could not see the back of the boat through the wind-driven water, spray, and hail. People's coolers and jackets had begun to block the scuppers and water was building up on deck. You could not stand up, so I crawled on hands and knees to the offending items and removed them. As soon as I lifted them, they were torn from my hands never to be seen again. People were crying, some were praying, others moaning, and then the wind increased yet again. I was still back beyond the end of the canopy near the engine trying to remove things so the deck would drain. Even over the shriek of the wind I could hear the wail of the engine, which must have been operating at maximum output. I expected to feel the sickening thud of wood on ledge any second, as I assumed the anchor had either pulled out of the bottom or the line had parted.

These conditions seemed to last for hours, but, probably went on for no more than ten minutes before the wind abated as quickly as it started. Ten minutes after that, the rain ended and the wind reduced to a moderate breeze from the west. The squall was over, and the *Lady Bev* and her passengers had survived. I went forward to the wheelhouse to report that no passengers had been lost overboard, and other than some clothes and coolers that had blown away as I cleared the scuppers, everything was in order. Dick instructed me to haul in the anchor line. There was very little weight, and when I grabbed the shank of the anchor to lift it in, the rest of the anchor was gone. The anchor had been torn apart by the force of the wind in the storm.

About this time, the radio came alive with maydays and calls for vessels requesting assistance. The sailboat I had seen had been flipped on its side and three passengers were missing. Numerous recreational boats were awash and several more people were missing from them. One party boat had dragged anchor and the engine could not turn the bow into the wind to lessen the force. The anchor caught in a rock crevasse and the boat ended up in six feet of water on the edge of the ledge. Another party boat with a cabin reported the roof of the cabin had been torn away and was lost. Those boats that could joined the search for missing mariners. Some were found swimming, some washed up on islands, and others dead. At the end of the day, five people were missing and ultimately declared dead.

The fleet limped back to their various harbors and learned several facts about this weather event. A tornado had been reported in Salisbury Beach, Massachusetts, killing several people. Salisbury is roughly fifteen miles to the west-southwest of the islands, and that was the direction the storm came from. The anemometer (wind gauge) on White Island, which was two hundred feet from our position, reported sustained winds of 130 miles per hour before the wind gauge blew away. The wind probably blew harder, but the weather service could not offer a number. The passengers reached shore soaking wet and minus a few items they had arrived with, but alive with a sea story I am sure they never forgot.

The second incident of the year occurred on a Saturday in the third week of September. I entered Needham High School in the fall of 1968, but I would work weekends until Columbus Day, when the party boat season ended. A hurricane had passed by well offshore, and although the weather in New Hampshire was sunny and calm, the offshore storm had generated a very large roll, bringing big, long-period swells in from the east-southeast. The fleet made the decision to sail.

I am not an expert in wave dynamics, but basically when a big, long-period swell encounters rapidly shoaling waters, the wave dramatically steepens and, if it steepens enough, the top pitches forward and breaks, like what you see in the surf on the beach. Waves come in series of three and seven. These waves may be as much as a third higher than the waves in between. If a vessel should encounter a breaking wave, there is the possibility of severe damage to the vessel's structure as the wave breaks over it or the possibility of hitting the bottom as you come down the back side of the wave. Hampton Harbor has a shallow water sandbar that runs for about a quarter mile before the water deepens off. Basically, if the sea state is reported as eleven feet or higher, do not bother trying to come or go, because you will encounter breaking white-water conditions. However, during this period, weather buoys did not exist, and sea state was estimated from shore.

Once you commit to crossing a bad bar, you cross it. Should you try to turn around and the wave catches you broadside, you can be rolled right over. Before we left, Captain Dick had given the passengers instructions about staying seated until I came back to give the all clear and to stay spaced out evenly around the vessel for stability. The two boats ahead of us took a little thrashing, but encountered no breakers, clearing the harbor without incident. The pilothouse on the *Lady Bev* was about ten feet aft of the bow and had a passage out on each side that accessed the deck. A window could be opened outward in the front of the wheelhouse for better visibility. There were no doors. About halfway across that quarter mile of shallow water, Dick muttered an obscenity. We could both clearly see a wave much higher than the others approaching the eastern edge of the bar. Dick gunned the throttle to cross as much of the shallow water bar as possible, but it was apparent the wave would crest on the bar before we cleared it. What happened next took a matter of seconds. Dick pulled the throttle back to an idle and turned the bow to take the wave head on. The wave began to crest with a height about ten feet above the top of the wheelhouse. As the boat hit the breaking wave the boat shot up in the air reaching a nearly vertical angle. The wave broke over the wheelhouse and through the window, filling the wheelhouse with water almost to the overhead. Just when I thought I might drown, a hand lifted me and my head above the water. Dick had picked me up with his left hand like I was a toy while steering with his right. The water drained away as fast as it came in while the *Lady Bev* crested the wave and plunged down the other side.

What I did not learn until later was, Dick could clearly see the bottom on the backside of the wave. However, the wave had so much momentum it stopped the boat's forward progress, even pushing it back some, allowing water to fill in behind the onrushing sea. If that had not occurred, the *Lady Bev* would have hurdled down the backside of the wave and plowed bow first into the bottom, likely shattering it to pieces. After idling to deeper water, we made a safety check of all compartments, and finding no damage, continued to the fishing grounds. Luckily, the swell subsided during the day and the return trip was uneventful.

High School in Mind

High school was much more pleasant than junior high. The classes were also more tailored to specific subjects. I took biology that year and not only did I enjoy it, I found it interesting and stimulating. Students were already beginning to meet with guidance counselors over college selection and field of major, and I chose biology, both because it seemed understandable and logical and because I could tailor the subject matter to my interest in fishing.

In the middle of the winter, I received a call from Bill Eastman, who spent his winters in Florida, wanting to know if I wished to be his mate next year. His crewman had entered college and would not be back in time for the start of the season. I jumped at the chance for two reasons. First, because his boat ran every day with the walk-on trade and never took charters, and second, because he was the best mackerel fishermen I knew. I could now make enough money to start saving to buy a car when I obtained my driver's license.

Two items had to be addressed before I decamped to Seabrook for the summer. First was finding my own place to stay. Lester and Marcia's kids were getting older and needed what limited space was available in their house for themselves. Bill Jr. and his wife, Anne, had purchased a travel trailer they stayed in behind Eastman's larger parking lot when the Hampton Beach traffic was too jammed up to go home to their house. Their kids were growing, too, and they had purchased a single-wide trailer to stay in for the summer. The travel trailer had two bunks, a two-burner stove, a small refrigerator, and a tiny toilet hooked to a septic system of dubious quality and legality, but it was my home for the next six summers.

Over the course of this winter, I went through my teenage growth spurt. I grew nine inches in six months to almost six feet tall. I felt like the cartoon character Baby Huey, who in cartoons sprouted full grown appendages in a matter of minutes. I was skinny as a rail, but hauling the boats anchor up to twelve times per day put some muscle on the arms, legs, and shoulders. On the *Lady*

Lil II, you stood on a pair of heavy timbers about a foot off the deck, bracing your knees against the rail to haul the anchor. This gave you a better angle both to see the direction of the rope and to give the captain hand signals to correctly position the boat. One day, while engaged in this process, someone picked my wallet out of my pocket. I was too young to have any papers or credit cards and the amount of money was negligible, but it was the idea. What kind of lowlife picks a kid's pocket? I told Captain Bill, and he spent the rest of the afternoon scanning the customers for the thief but never located the guilty party. This was probably just as well, as the beating that would have ensued would have been awful.

As the summer progressed, Bill began to train me to be a captain and in the art of finding fish. I learned how the mackerel moved from offshore in the spring to the coastal ledges in June. Further, he taught me how the fish moved on and off the shore with the incoming and outgoing tide and then later in the summer over to the Isles of Shoals.

In the privacy of his tiny wheelhouse, roughly the size of a telephone booth, he directed me to pick the spot where we should move to fish. One day he turned and said, "Ok take me to that spot." I was both elated and terrified. What if I couldn't find the ledge? What if we didn't catch anything? I took over the wheel, checked the compass heading, and checked the time on my watch. This was to calculate how long to run. I knew the speed of the boat was nine knots and the distance was about two miles, so roughly ten minutes of steaming. Once we got close to the destination. I could use landmarks to find the top of the ledge.

You must remember in this period, there was no GPS or LORAN (shorthand for LOng-Range Aid to Navigation) and only a few boats had radar. You navigated plotting courses on a paper chart run off on a magnetic compass and calculated distance by knowing the boat's speed and using a watch or chronometer to run a specific time. Landmarks are a form of triangulation. Pick two sets of two fixed objects, one in front of the other. As you run a distance, the object in front will move from one side of the object further in the distance to a position where they align to moving to the other side of the object. When both sets of objects you have selected align, you are at the point of a triangle and can be nowhere else. For example, align the water tower in Hampton Falls with the draw span house on the Hampton Bridge and the church steeple in

Another spring in the boatyard. David leaning against the *Lady Lil II* just prior to launching for the season, Newburyport, MA, 1976.

Hampton Falls with the left leg of the water tower on Hampton Beach and you will be on top of an underwater ledge three miles east of Hampton. Landmarks are more accurate than GPS because GPS has a slight error built in to prevent terrorists or opposing militaries from exactly fixing targets in the United States.

I lined up the marks and Bill took over while I laid down mackerel chum and tossed the anchor. Then we waited. Nothing happened for about five minutes, not a bent pole, nothing, and I feared I had failed my first big test. Bill always fished a spot for twenty minutes, as fishing requires patience and the fish need to follow the chum to the boat. Suddenly a rod bent, then another, and another, and finally every rod on the boat. The mackerel stormed aboard as people yelled and shouted as they filled their pails and coolers. I was quietly elated, as I had passed my first test with flying colors.

CHAPTER 4

Fast Cars and Freedom

I returned to Needham and entered high school as a junior. After taxes, I was over half-way to a new car. The winter passed very slowly. After school, I listened to reel-to-reel tapes I had made of music I liked and had a multiband transistor radio tuned to a frequency fishermen conversed on. My father had met a new love interest on the boats that summer, a single woman named Alice who had three kids, and soon he moved to Westborough, Massachusetts, to live near her. His behavior was becoming increasing erratic, fueled by liquor, cigarettes, and tranquillizers. This new love interest sent my mother into a tailspin, as I think she finally realized he was gone for good. She, too, dove into the bottle and swam around like a fish in a pond.

I was invited to work on getting the party boats ready for the spring season. This would provide extra income and further my knowledge of maintaining wooden boats. I just needed a plan to get there. Financially, I had done well the previous season, both because the mate's pay had increased and I had begun selling items on the boat like mackerel jigs and plastic bags for people to take their fish home. Every Friday, I would place essentials in one of the plastic bags I sold and leave the bag in my school locker. I would walk from school to the train station in town and take a train into South Station in Boston. Then I would ride the MBTA to North Station and take a train to Newburyport where someone would pick me up. While waiting for the train to leave for Newburyport one day, the engineer walked through the cabin. I recognized him as a man who had a boat hauled in the same boatyard that the Eastmans used. I said hello, and he recognized me and invited me to ride with him up front. The train was called a Budd liner, a form of self-contained engine and passenger car. The tracks to Newburyport were in terrible condition and the train had a number of stops. However, the final seven or so miles were a straight-away across the marshes of Newbury. The train here reached speeds of more than eighty miles per hour, bouncing and rocking violently. I asked if he ever worried about derailing. He just grinned and said maybe then they would repair the tracks.

My closest friend was a young man who also worked on one of the boats and was in his first year at Northeastern University. He often picked me up at the station. We both spent the weekends in March and April working in the boatyard, sanding and painting our respective boats while also learning how to conduct yearly maintenance like repacking stuffing boxes and caulking garboards. The garboard is the seam between the first plank on the bottom and the keel. To caulk it, you needed a wooden mallet, a caulking iron, and rolls of cotton. You then took the cotton and tapped it in the seam gently with the iron so it would stay in place while laying on your back in the mud. You then went back with the iron and drove the cotton in the seam with the mallet until it was tight. The seam was painted with red lead and covered with seaming compound. When the boat was placed in the water, the wood would swell, and, if you had done your job well, there would be no leaks. The reason I describe this is because it is one of the arts that had been passed down for centuries but is now lost because boats are now almost exclusively made of fiberglass or steel and caulking is no longer required.

I passed the driver examination in late spring. Looking back, sixteen and a half seems awfully young to be driving, but, in a show of support, my father would occasionally let me drive his car to Seabrook. He always owned Ford station wagons, which were enormous. It seemed like you needed binoculars to see over the hood, but there I was bombing up Interstate 495 on my own. Somehow, through luck more than skill, I never was in an accident.

One other notable event occurred that spring. My friend John stood in line at the Boston Garden to buy tickets to the Stanley Cup playoffs. The Bruins were the hottest team in sports, led by Bobby Orr, who, I would argue, is the greatest hockey player of all time. The tickets he bought for us were standing-room in the second balcony, but the ushers were mostly reasonable and would allow you to kneel behind the seats in the first balcony. We saw most of the games, including the final one. That day was incredibly hot. The old Boston Garden was a filthy, rat-infested, steam box. The temperature in the second balcony was well over one hundred degrees and the Garden ran out of beer, an unthinkable sin in Boston. Some patrons left both for beer and air-conditioned bars across the street. We snuck down into the loge seats near the goal the Bruins would be shooting at and sat down in a pair of empty seats. The ushers down there would toss you in a second and the paying customers usually called attention to seat jumpers quickly, but the people we sat next to knew the people who had left were not coming back. The game commenced, and as every hockey fan knows, Bobby Orr scored the winning goal a little over forty seconds into overtime. He flew through the air, part in jubilation and in part because the Saint Louis defenseman had tried to clean him out. The cameras clicked and the most iconic picture in Boston sport's history was published. The place erupted and the people next to us broke out nips and champagne they had smuggled in, which they shared with us. The bedlam inside was exceeded by the pandemonium outside. By the time we exited the Garden, people were buying cases of beer and handing them out on the street. I have never seen any celebration like that since. If you look very carefully at an uncropped picture with a magnifying glass, you can vaguely make out a blurry image of a kid with glasses next to a skinny blond youth. That would be John and me.

I continued to hone my skills. I was now allowed sometimes to dock the boat or run it out of the harbor under less-than-ideal conditions. Bill had purchased a new used boat that spring from Montauk, New York, which he also named *Lady Bev* and the original *Lady Bev* was sold. The boat was just under sixty-five feet in length, held more passengers, and was much faster. This was the second new upgrade for the fleet, as the World War II, converted navy motor launches were reaching the end of their serviceable life. Pay was increased again because more people meant more work. I would have enough for a car in the fall.

This was also the year I discovered the fairer sex. A lot of parents took their entire family fishing, and sometimes this included girls around my age. The girls fell into three categories: those who truly enjoyed fishing, those who treated fish like they had some form of leprosy, and those who just wanted to be as far away from their hopelessly square parents as possible. My job was to teach people how to fish, and sometimes the catch was more than mackerel. These were all summer romances, walking on the boardwalk on Hampton Beach or attending a bonfire on Seabrook Beach. They ended when the vacation ended, and you never saw or heard from the girls again. I did, however, meet one girl this year who exchanged addresses at the end and promised to write. We kept up correspondence for three years even though we never met again. I often wonder what is lost in this modern age of texts, ghosting, and selfies compared to receiving a carefully constructed letter on scented paper. Personally, I would take the letter over texts any day of the week.

That fall, my father promised to take me to buy a car after my birthday. Even though it was my money and my car, I needed someone over eighteen to execute the contract in Massachusetts. We went to a Ford dealership in Westboro, and I selected a 1970 light blue Grand Torino GT with a 351 engine, two-barrel carburetor, and blue racing stripe. I received a discount for buying a 1970 model when the 1971 models were already out. The two-barrel carburetor was a necessity as I could not obtain insurance for a four barrel, which was considered a high-performance engine. My father signed the papers and sold it back to me for a dollar. After insurance, sales tax, and registration, I still had money left over for gas. I was finally free to come and go as I pleased . . . almost.

My mother immediately tagged me to do the grocery shopping and transport my brother here and there. I was not as free as I thought. I obtained a senior parking pass and drove to school. No more bus for me. Word spread fast, not only about the car, but about how I owned it. I went from one of the countless masses to somebody. This was both amusing and depressing.

As May rolled around, thoughts turned both to college admissions and graduation. I had applied to four colleges to major in biology. I was accepted to three and wait-listed at UNH, which had been my first choice. UNH would consider wait-listed people after they learned how many New Hampshire residents accepted. This was a gamble I could not afford. I accepted the invitation to Boston University. Graduation occurred in early June and was to be held on the football field. A thunderstorm had other ideas, and the assembled multitude broke into a sprint up a steep hill to the school gymnasium. Note to self: do not try to run in a gown! The soggy ceremony concluded; I was officially free of high school. My mother had invited Bill and Lil Eastman to come to her house for cocktails and hors d'oeuvres. I was surprised they accepted, as Bill usually was in bed by eight and graduation was not over until seven. They had never met before, and I was worried about my mother's behavior, but the evening concluded without a hitch. I attended the town graduation party, which lasted all night. Once you were in you were not allowed to leave, as countless high school seniors are killed in drunken-driving accidents on graduation night. About midnight a girl who I barely knew approached me about leaving and wanting to know if I had any beer. We told the chaperone at one of the doors we were stepping out to get some air because the building was stifling, which was true. We would be back, which was also true, but not for four hours. We spent the time sipping and talking about our aspirations and what we would do with our lives beyond high school while listening to the car radio. We returned about the time breakfast was being served. I never saw or heard from the girl again.

CHAPTER 5

College Years:
Between the Whiskey and the Novocain

Lying in my hospital bed, sifting through decades of living, it became evident the most jam-packed and ultimately jumbled memories in my mind were my college years, which began in the fall of 1971. Sure, my massive head trauma was partially to blame for any confusion, but so were the times, which were as confusing as any ever.

The year 1971 was a tumultuous one. The country was deeply divided over the Vietnam War. Boston especially was a hot bed of the antiwar movement. Kent State had solidified everyone's positions a year earlier. Yes, they really will shoot you for voicing your opinions. Many student antiwar groups that espoused nonviolence had turned violent. The cops did not like the kids and the kids did not trust the cops. Hampton Beach has always had a big police presence to keep order. With often over one hundred thousand people crammed into two miles of sand and boardwalk, one quick flash could set off a riot, and my summer between high school and college saw a doozy.

The Hampton Beach Casino is a large music venue at the center of Hampton Beach built in the 1890s. Being of wood construction, the fire codes are strictly enforced, including the number of people allowed in the building. The casino once favored big bands with ballroom dancing, but the times they were a'changing, and more and more rock acts were booked. On July 8, Jethro Tull, an up-and-coming rock act, was to take the stage. Whether purposely or by mistake, the casino sold at least four hundred more tickets than would be allowed in. The fire marshal closed the entrance, and the people holding tickets that were turned away went nuts. Kids tried to storm the doors, the cops fired gas and began busting heads, and the riot was on. People poured into the streets and tried to come onto the beach to join the riots, especially from 1A to the south. The Hampton Harbor draw bridge was raised to keep rioters from Massachusetts from getting to the beach. Of course, now people who wanted to leave could not get out. The people who wanted to join the riot milled around in the stalled traffic on the Seabrook side of the bridge. Some tried to swim across the river. About 10:00 p.m., four teenagers walked down Eastman's pier towards the boats. Bill and Lil had been watching the boats nervously, waiting for trouble. Bill retrieved a revolver he kept in his bedside drawer and headed down the stairs. I was told to stay put. As Bill approached them, the ringleader told Bill loudly, "Hey, old man, these boats are going to take us to the riot." Bill never said a word, but when the kid opened his mouth again, Bill pulled the revolver from his pocket and shoved it in the kid's throat. His eyes went wide, and their hands all rose slowly. Bill marched them all up the pier, told them to beat it, and if he saw them again, he would just shoot them. They did not return.

The summer passed in anticipation. I was now quite confident in running the boat, and since Bill's eyesight was declining, he often had me run the boat in the fog. He wore bifocals and when the mist covered his glasses, he had a hard time seeing. I also noticed for the first time that he forgot things occasionally. This was a man with a photographic memory. This was nothing serious, just different. A couple weeks after the riot, Bill had a terrible toothache. He called his wife on the citizens band radio to see if the dentist could see him over the noon hour. I drove him to Newburyport to a dentist who looked to be about eighty. Bill told him he did not have time to waste and to just pull the tooth because he had an afternoon trip to run. Two shots of Novocain were administered. The dentist kneeled on his chest and using a pair of pliers vigorously yanked on the tooth while bending it back and forth. There was blood everywhere and the tooth appeared to be holding strong. Finally, in one last burst of strength, he tore the tooth out with a popping sound and it landed on the tray. Bill paid the man and I drove him back to Seabrook with an ice pack on his face. Upon arrival, he went upstairs, had two double shots of whiskey, replaced the ice pack, and staggered down to run the afternoon trip. The second mate would handle the deck and the lines, I would run the boat, and Bill would sleep in the back of the wheelhouse. Luckily the weather was good, the fishing acceptable, and we pulled it off. Bill stood up a couple times so people could see him, but between the whiskey and the Novocain, he was out for most of the afternoon.

Around this time, I met my first serious girlfriend. All the kids on Seabrook Beach hung out together. Ages ranged from about fourteen to twenty. The drinking age was still twenty-one, so hanging out on the beach was the entertainment plan. I met Kristine at a bonfire one evening. She had an enigmatic smile which caught my attention. We chatted, then broke away from the crowd, finally leaving in my car to talk privately. She was fifteen, and I was seventeen. Our bond was teenage angst which kept us together on and off for five years.

Fall of 1971: College Begins

I had done especially well on the biology achievement test, scoring a 781 out of 800, and because of that, in college I placed out of the two semesters of first-year freshman biology. I was supposed to enroll in sophomore genetics, but the course was full, so I took a night course in genetics with adults in continuing education. This really fouled up my schedule because I had to stay at school two days a week for a course that began at eight at night. Similarly, I had to take four semesters of French because I scored spectacularly poorly on the French achievement test. BU was a liberal arts college and required two courses each in political science, humanities, and social sciences. I had no interest in these subjects whatsoever, but compared to the science and math courses, they were easy A's. I took one every semester for the first three years.

As a student, I couldn't afford to park in Boston, so I took the MBTA green line trolleys from Riverside in Newton, where you could park for free. The trolleys were relics from the early 1900s: decrepit, disgusting, dirty, and prone to breakdowns and fires. The ride on a good day was an hour, and I got off at Fenway Station to save the quarter that riding one more station would cost. I then walked about ten blocks through a dicey section of Boston to BU. BU was an urban campus laid out along Commonwealth Avenue from Kenmore Square to the Alton neighborhood of Boston. Commonwealth Avenue is a six-lane highway with another branch of the green line running down the middle. The function of every Boston driver is to get from zero to sixty between each set of redlights. Pedestrians are fair game, and the trolleys operate by their own set of rules. Crossing Commonwealth Avenue to go from class to class was a life-challenging experience.

College was not particularly hard, just different. Nobody stood over you taking attendance like in high school. If you wanted to apply yourself, you could do well; if you wanted to major in beer drinking, you could do that, too.

Since I was paying for tuition out of my fishing money, I figured I may as well get my money's worth.

When I turned eighteen that fall, I was required to register for the draft and obtain a draft card. The local draft board was in Wellesley, Massachusetts. You had to show up in person, answer a bunch of questions, and get a quick once over. Basically, if you could stand up, were not married with kids, and could walk across the room, you were classified 1-H, which stood for "holding," that is, fit for duty. Your classification would change to 1-A if you were drafted. The draft system had recently been changed to make it less biased. Education defer-ments had been eliminated, and a lottery system was put in place which assigned every calendar day a number. With the war still raging, experts predicted more than half the numbers would be called.

In 1972, the voting and drinking age had been lowered from twenty-one to eighteen. The lottery was shown on television and kids gathered around, alcohol in tow. Those whose birth numbers were called early drank to drown their sorrow, those in the middle to drown their uncertainty, and those with high numbers to their good luck. However you sliced the pie, I noticed considerably fewer people in class the next day.

As part of this same legislation, several federal age requirements were low-ered. Among them was the year you could sit for the US Coast Guard test to captain a party boat, technically called an ocean operator permit. The test had many mandatory requirements, including documenting sea time served under a licensed mariner. This document had to be signed by the mariner or mariners in front of a notary. I had this document with days at sea to spare as well as a completed physical and eye and ear test. The test was given on Thursday every week at Coast Guard First District Headquarters on Atlantic Avenue in Boston. I sat for the test the week before Thanksgiving in 1972.

Once your paperwork was approved, you were provided the exam ques-tions "50 Rules of the Road," on which you had to score 90 percent or higher, a navigation problem which you had to get correct, and 100 general mariner questions which you had to score 75 percent or higher. The test was multiple choice, but you had to read the answers very carefully, as they often only dif-fered by one word. Most people did not pass their first try, but I scored a 94 in the rules of the road, an 80 in the general, and aced the navigation problem.

There was one last possibility for the Coast Guard to deny you a license, as the fine print says the license is issued at the discretion of the Officer in Charge of Marine Inspection (OCMI). I was ushered in to see a high-ranking non-commissioned officer who looked to be in his mid-forties, probably with over twenty years in the Coast Guard. He asked me a few navigation light and day marker questions, which I answered correctly. Finally, he asked what does the light combination red over white over red over red signify? For a moment I froze because I did not know of such a light combination. Red over white over red signifies a vessel restricted in its ability to maneuver. So, I led with that followed by a statement that there was no such light combination, unless of course it was a Coast Guard cutter backed up to a whore house. My patience with this game of cat and mouse had grown thin, and my sarcasm was showing. Sensing his last attempt to trick me had not worked, he threw in the towel with, "Damn, I do not think any nineteen-year-old kid should be granted a license, but I cannot find a reason to deny this, so do not screw it up." I left just after dark with my license in hand.

I was immediately offered a job for the following summer running the oldest boat in Eastman's fleet, the *Lady Lil*. This was a good thing because BU had hired John Silber from Texas to radically improve the BU financial position and their academic standing. This would take money and lots of it. Next year's tuition would be double this year's. Boston University ended classes the last week of April. President Nixon, while promising to end the Vietnam war, had just ordered a massive bombing campaign in Laos and Cambodia as well as North Vietnam. While one could argue this was to increase his bargaining position at the never-ending peace talks, the anti-war movement felt betrayed. Violence broke out across the country on a massive scale, including at BU. BU had a week for studying for finals before tests began in early May. Massive protests began daily. The students would mass at the west end of Commonwealth Avenue and the tactical police squad would mass in Kenmore Square. When they met, the scale of the violence shocked the nation. The students hurled ballast rocks from the trolley tracks, trash cans, and anything else they could lay their hands on at the cops. The cops used night sticks and fired tear gas as well as unleashing dogs on the students. After five straight days of street battles, BU canceled finals and sent everyone home. Thus ended my first year of college.

My sophomore year of 1972–1973 passed uneventfully. Kristine and I settled into a routine of hanging out with her friends on weekends in the winter and going on road trips exploring New England with my friends in Seabrook in the late summer and early fall. One of the crewmen, a very jovial young man named Richard, had a giant LTD with an enormous V-8 that was ideal for these trips. Six of us would pile in with a case of beer in the trunk and set off to explore state parks, lighthouses, and whatever caught our fancy.

In the winter, I would drive to Seabrook and hang out with Lester Eastman. Lester was a hard worker and outstanding fisherman, but he liked the bars and had a whole bevy of good-time buddies. From the time I got my driver's license, I was the getaway driver for this band of characters. Salisbury, Massachusetts, borders Seabrook, New Hampshire, and, at that time, had more bars per capita than any place I knew. Lester could find trouble in just about any one of these establishments. One of his favorite phrases was, "Me and the kid will take on any two guys in the house." My response was, "The kid is leaving, are you staying or coming?"

The winter activity of choice was playing pool at a club on Salisbury Beach. Mostly people hustled each other for drinks, but occasionally there was serious money laid down. The cast of characters looked like the extras from a *Star Wars* movie. Several had noticeable bulges under their arms or back pockets indicating they were carrying pistols. This was not a place where you misspoke or accidently spilled a drink on someone. This was, however, a good place to sharpen your street smarts and negotiation skills with people who have had too much to drink. Lester was a good pool player when reasonably sober, and I was adequate but streaky. We usually drank for free until Lester had consumed too much alcohol. I would offer him a ride home, which Lester seldom took, but somehow, he seemed to always get home eventually to his long-suffering wife.

BU had settled into its familiar spring pattern as well with riots resuming at the end of April. The spring of '73 was different. The new university president John Silber, who'd started in January of 1971, was a distinctly law-and-order man from Texas. Apparently having had enough, he announced that in the spring of 1973, no matter what, finals would not be canceled and those who failed to take their exams would receive a zero, effectively flunking them out of college. Professors who canceled finals or failed to show up would be fired.

While I personally welcomed the announcement, it presented me with logistical challenges. As a commuter, I had to get into the buildings before the daily scrum ensued. Luckily two of my finals were at 8:00 a.m. and the violence did not usually get going until about ten, but one other final was at eleven and the last one at 1:00 p.m. So, on each of four different days, I had to be in the building before the trouble started.

One day stands out. I arrived for my calculus final and tried to enter the building. This required climbing about ten granite stairs flanked on each side by yew bushes standing about three feet higher than the stairs themselves. On the flight of stairs was a skinny, long-haired gentleman I vaguely recognized from my calculus class. A quick clash ensued. He informed me the students were on strike and I could not enter the building. I informed him daddy was not paying my tuition, I was, and I was taking a final. He moved to block me, and I quickly stuck one arm between his legs, the other behind his back, turned him sideways, carried him to the yew bush and stuffed him headfirst into the tree. Then I entered the building and proceeded to the designated room. During the final, he appeared again at the classroom door, so he must have extricated himself from the bush. Yelling power to the people or some such foolishness, he tried to gain access to the room. The calculus professor gently but firmly nudged him out the door and locked it. I never saw him again; I guess he flunked out.

I readied the *Lady Lil* for the water in April and carried my first charter as a captain in early May of 1973. Charters are groups of people who rent the boat for the day to go fishing. Mostly they were groups of factory workers, social clubs, or bachelor parties. They received a discounted rate because they paid a flat fee even if only a few of their group showed up. They often rented a bus and drank on the bus ride. We used to joke about how long the day was going to be when the first man out of the bus fell flat on his face and the next thirty used him for a doormat. Every group had a core of people who truly wanted to fish. You tried to work with those people and hoped they could keep the rest in line. Usually, I gave a speech at the beginning about how the Coast Guard really frowns on two things: one is mutiny, and the other is coming back with fewer people than we left with, so stay in the boat.

My first career charter was typical of the genre. I had hired a local boy as my crew. He was fifteen but had black hair and a heavy five o'clock shadow which made him look older than me. We loaded up our charges and their beer and headed offshore. The day went well, and the fishing was good, but by about one o'clock some of the passengers were getting drunk. I had the mate pull the anchor at two, told the charter we had wooden boards to clean their fish on if they wished, and headed for home. About ten minutes later, one of the leaders of the group came up and said we had a situation. I idled the boat down, grabbed a club I kept in the wheelhouse, and went back to survey my options.

What I learned was the crewman had asked a drunk who was not using a cutting board to use one. This was a reasonable request, but the drunk took exception, spun around and held his hunting knife to the crewman's neck. The rest of the charter was trying to talk him down, but they were not succeeding. My first thought was, are you kidding me, on my first day? I went down the port rail with the club hidden against my right leg. The drunk was focused on the knot of people in front of him trying to get him to put down the knife. I went to the stern, cut across to the starboard rail, came up behind the drunk, and unleashed a ferocious blow across both his lower legs with the club. He went down like a stone and luckily dropped the knife. Men from the charter subdued him and we tied him to a post for the ride home.

By the time we got in the harbor, the drunk was crying and begging not to be arrested. I cut him loose at the pier, told the charter they were welcome again, but not the drunk. On the way in, the men had taken up a collection for a tip for the crewman, which was a nice gesture, but he was badly shaken. At least drunks were predictable, and you could usually find a way to deal with them, but behavior like these foreshadowed dark days in the future as elements of this country ramped up illegal drug use, which often produced very dangerous, unpredictable outcomes.

CHAPTER 6

Bad Habits and Explosive Events

Fishing in 1973 was notable for what did not occur. The mackerel left in the middle of June and by July 1 no one was catching any kind of fish on the inshore ledges. No one knew how to respond. Even people with long memories had never seen this happen. All the boats switched to trying to catch bottom-dwelling fish such as cod and haddock, but even these seemed to have disappeared. All the captains became nervous as business began to drop off rapidly.

The customers that came gave younger captains like me a particularly hard time. Older captains, like Bill, got somewhat of a free pass by telling stories and pointing out that people had fished with him for years and always caught fish. I grew a mustache to look older and a group of us, both captains and crew, started smoking pipes. Hey, it worked for Popeye! Several crewmen shared a small shack where they lived on the edge of the parking lot. After trying to buy some pipe tobacco at the convenience store at the top of the street and finding the store was sold out, I stopped by their place to borrow some tobacco. There was the proverbial Prince Albert in a can on the bedside table. I filled my pipe and headed to my trailer. This was the grossest smelling pipe tobacco I ever tried, and soon I noticed I was getting dizzy. I finished my pipe on my porch and turned in for bed. While trying to fall asleep, I began seeing things that were not there, like mackerel with legs and other hallucinations before falling asleep.

The next morning, I found the two crewmen and told them their pipe tobacco was spoiled and should be discarded. The older one, who was in college, motioned me out of hearing range and had a strange smirk on his face. I was informed he kept marijuana in the can so people would think it was pipe tobacco. He thought the whole episode was funny, but I was not amused. That was my first and last experience with dope. I have been offered it more times than I can count over the years, but my feeling was, is, and continues to be anything that creates that much distortion in the brain cannot be good for you.

Late in the summer, the fleet discovered the reason for the abysmal fishing. One captain, who was using herring for bait instead of the usual clams and fishing for rock cod relatively close to the Isles of Shoals, had a customer hook a fish while reeling his line to the surface. The fish fought aggressively and when finally brought to the surface, turned out to be about fifteen pounds with a mouth full of razor-sharp teeth. Bill Eastman was contacted as the reigning fish expert. He also owned a copy of *Fishes of the Gulf of Maine*, by Henry Bigelow and William Schroeder. This was a textbook published in the 1950s by the U.S. Fish and Wildlife Service, which oversaw commercial fisheries at the time. The book is still considered a giant in the field, both for the information it contains and for the conversational prose used by Bigelow to describe the fish. Bill read a chapter every night to learn more about his lifelong passion. The fish was identified as a bluefish, common off southern New England and New Jersey, but only seen sporadically in the northern Gulf of Maine.

Calls were made to New Jersey, where many of the newer boats in the fleet had been procured, to find out how to rig up and fish for these fish. We learned you needed wire leaders for two reasons: first, because the hooked fish could bite through monofilament, and, second, because a trailing unhooked fish would strike at the bubbles released by the taught line slashing through the water. The fish traveled in schools and fed in a frenzy, so nearly everyone in the boat might hook a fish at the same time, unleashing a pandemonium of tangled lines. Finally, the teeth were lethal. The fish should be gaffed, held at a distance while still on the gaff, and the hook removed with a good-sized pair of long-nosed pliers. The impatient angler could easily be rewarded with a lost finger, lost toe, or severe laceration. These were not fish to be trifled with! Business immediately perked up. The fish were streaky, with some trips without a bite and others boating upwards of a hundred fish. The bluefish stayed until the end of the season and business remained strong.

The bottles of Maalox and other antacids were put on the shelf, but many of the captains had picked up a new bad habit, cigarette smoking. I returned to college in the fall of '73 with a pack-a-day habit, the weapons of choice, menthol. However, by then, the evidence on smoking was overwhelming. Gone from advertising were the tableaus of smoking being good for your health, calming your nerves, promoting good digestion and even better breathing that had dominated media in the 1960s. As a biologist, I could not countenance what I was doing to my body with the facts at hand, so I quit.

I ran charters again that summer of '74 and slept in the travel trailer at the rear of Eastman's parking lot. Sometime in August, while sound asleep in the middle of the night, I was awakened by a huge boom. I got up, looked around, did not see anything unusual, and went back to sleep. The next morning, I was awakened to swarms of people with guns, badges, and blue jackets with an assortment of yellow letters like ATF and FBI. The cause of the commotion soon became apparent. Our competitor's newest boat, the *Jubilee*, was very low in the water but still afloat. You could see wires running to the shore. Apparently, someone had set two charges of dynamite in two different watertight compartments, but one had failed to detonate. The one that did detonate sprung some planks and flooded one compartment. If both had detonated, the boat would have been on the bottom.

Nothing like this had happened before. Fishermen had beefs and occasionally mooring lines got cut, but this was on a whole other level. The ATF was tasked with finding the source of the dynamite, and the FBI was tasked with finding the bomber. Numerous other state and local forces were involved as well. Two brothers owned the business and were known to have had some passionate disagreements, sometimes involving clamming forks, oars, and other instruments of potential battering. The rumor mill went into overdrive laying out plausible cases why each brother could be responsible. The rest of us rowed out to nervously check our boats. After all, the Seabrook Post Office and a courthouse in Newburyport had been blown up by some crack-pot anti-war radicals. Maybe someone was trying out new tactics for upcoming events. You never knew.

A few months later, an arrest was announced. A local had overheard one of the brothers bitching about the other brother over his beer in a bar. The two never made contact and the local was not hired. He just did it! Hopefully everyone in the area learned two important lessons: loose lips really do sink ships, and do not gossip about people until all the facts are presented. You are usually wrong and can ruin someone's life with false inuendo.

So You Want to Be a Biologist

September of 1974 saw a return to Boston University for my final year. Having placed out of freshman biology, I had exhausted the undergraduate courses by the start of my senior year and was invited to take graduate-level courses such as ichthyology (the study of fish) and oceanography, which covered a wide range of chemical, biological, and physical processes of the oceans. These courses were stimulating but intense. I spent some time in the BU library but most of my research was done across the Charles River in Harvard's Museum of Comparative Zoology's basement library. Here I discovered they still sold Bigelow's *Fishes of the Gulf of Maine,* of which I eagerly purchased a copy, which remains one of my most prized possessions. Several times a week, I rode to school with a fellow biology major from Needham named Pete. He had discovered an ingenious, if not entirely legal, way to beat Boston's notoriously high parking rates and equally zealous meter maids. In Boston, you cannot park in a parking space if the meter is missing or broken. One day, Pete discovered a meter that was loose in its cement casing and took it home. There, he took the meter apart and set it to read the maximum time. No one would park in the space because there was no meter. He would arrive, look around, take the meter from his trunk, and plant it in the ground. Pete had free parking for most of the winter until an enterprising meter maid must have figured out the needle never moved. We went to go home one afternoon, and both his car and the meter were gone. The car was in impoundment in South Boston and Pete had to pay three hundred dollars to get it back. We were back to riding the MBTA for the rest of our time at BU.

That year, my father was busy with his new family, and I seldom saw him. The exception was when he needed help moving, which was quite often. One weekend, he moved from one house in a subdivision in Westboro to the neighboring house. I have no idea why. The houses looked identical to me, but wife number two liked the house next door better, so I spent the weekend in a conga line of furniture that stretched across the front lawns from house to house. This would be the fourth of more than thirteen moves over the coming years.

His bad habits proceeded unabated and were on full display when I met him for lunch about once a month in downtown Boston. He was now chief counsel of the John Hancock insurance company in Boston. I am sure the company did not invent the three-martini lunch, but they certainly seemed to endorse it. We would meet in his office and walk to the Red Coach Grille where we ordered, you guessed it, martinis. Martinis are called silver bullets in the bar trade for a reason. They are basically glasses of pure alcohol which apparently only James

Bond could drink in large quantities without showing any ill effects. Lunch was always the same: a Red Coach sirloin burger with fries, the house specialty, with two silver bullets for me and three for my dad. We were not alone. All the big office buildings emptied out for lunch, and if you looked around no one was drinking diet coke. Manhattans, gimlets and martinis, and double bourbons on the rocks were the lunch fare. I always wondered how any work got done after lunch in Boston. I would return to BU for a two o'clock class. This year it was philosophy of science. Luckily, the lower-level classes here had enormous numbers of students and were given in lecture halls. I would take a seat towards the rear and usually would be deep in a power nap within half an hour.

I ended up writing several scientific papers that fall. My first was on bluefin tuna, using fathometer chart paper which showed what bluefin look like on a fish machine, and in one particularly striking sequence, a fish rising under the boat, grabbing a bait, and swimming down to escape. The professor was enthralled, and for the first time I realized that many PhD biologists have little or no real-world experience in the fisheries. I had worked as crew on bluefin trips for four falls at this point. The party boats would be used to catch bluefin for commercial sale to augment income when the tourist trade abated after Labor Day. Bluefin tuna follow edges of underwater ledges and banks that have steep contours. They are most likely to be encountered where these contours change directions, essentially forming an underwater corner. Bluefin are one of the largest fish in the sea and are capable of incredible speed. On an initial run, a bluefin may hit speeds of sixty miles per hour, or, put in perspective, over fifty feet per second.

The equipment used during this period was a specialized hook made in Japan attached to a cable wire leader attached to a nylon shocker, which, as the name implies, is to allow some stretch in the line so it does not break. This is all tied to a basket of 3/8-inch-diameter cotton rope containing approximately six hundred feet of line. The bottom end of the line has a pre-tied loop hanging over the edge of the basket so another basket of line can be tied on quickly if necessary. The lines are set at various depths and baited with things bluefin like to eat such as whiting, herring, or mud hake. You can use either live or dead bait. The tuna lines are tied to the rail with thin pieces of line that will break, making a noise and setting the hook, but weak enough not to break the tuna line. The baskets are soaked with water and the fishermen wear wet cotton gloves to keep their hands from being blistered by the line speeding through them. When a line goes off, the nearest person begins to play the fish by placing the line over his belt and hip to apply some pressure to tire the fish, but not enough to break the line. The other people on the boat, if there is more than one, pull in the remaining lines and tie a second basket of line on the line with the fish on, in case the fish cannot be stopped before the original line is pulled out of the boat.

The adrenaline rush is immense, and many fishermen become addicted to catching these majestic fish. Over my lifetime, I have seen far too many fishermen lose their boat, their house, or their marriage in this fishery. What ensues is a giant tug of war lasting from twenty minutes to four hours. Apply too much pressure and you break the line, apply too little and the fish never tires, and the hook eventually wears a hole in the fish's jaw and the hook falls out. Most initial runs take six hundred to nine hundred feet of line, with secondary runs taking less. When the fish begins to tire, it will switch from making runs to taking the line down deep and making slow circles under the boat. The goal at this point is to try not to let the fish take any line and to gain a few feet at a time on the far side of the circle when the angle of the line favors the person playing the fish. Eventually, you will reach the shocker and then the wire leader. At this point, the fish will be in sight, showing an awe-inspiring silver glow deep in the water. The person not playing the fish will then attempt to harpoon the fish right behind the pectoral fin to hopefully hit the heart. When the fish is harpooned, the dart head comes off the dart poll and is attached to another basket of line in case the fish makes a final run. If the person throwing the dart is accurate, the fish will pretty much give up, but if the person throwing the harpoon hits the head, the dart will bounce off; if he hits the back or well behind the pelvic fin, the fish will have a flesh wound and a renewed desire to escape.

Once the fish is brought all the way to the surface, the tail is lifted clear of the water and a stout line called a tail strap is applied over the tail and tied to a bit on the boat. The fish are still dangerous at this point. I was once nearly knocked unconscious by a wildly thrashing tail, and I saw a boat have a plank stove in, luckily above the water line, by an angry fish furiously beating its tail against the hull. Finally, the dart poll is inserted under the gill plate to cut the fish's gills so it will bleed out. The fish is then towed backward through the water by the boat to drown it and complete the killing process. After the captain is sure the fish is dead, he will load it into the boat for transportation to a tuna buyer.

The Japanese consider North Atlantic bluefin, which they call Boston bluefin, both a delicacy and a ceremonial necessity. In this period, the buyers would have a Japanese grading technician in residence who would grade the fish based on internal temperature, shape, and fat content, among other things. If the fish was deemed worthy of shipment to Japan, the fisherman would receive thirty to fifty cents per pound for the headed and gutted weight of the fish. Since the fish averaged five hundred to eight hundred pounds cleaned, this was a good day's work for the time period. Fish deemed not worthy of shipment would get five to ten cents per pound and be shipped to a cat-food cannery in Maryland. All this information and more on dietary habits, fishery locations, and how a fisherman could entice a wary fish to bite a line were included in my paper. The professor was most impressed, and I was awarded an A+.

My second paper was my first foray into fishery management. The thesis was basically, that the only form of management prior to World War II was, in fact, war itself. I showed that prolonged periods of peace allowed for overfishing to occur repeatedly through European and American history. Most people at this time still debated whether overfishing was even possible, and should they concede it was, only modern mechanized fishing could cause it. This simply was not true, as landing records from both Europe and North America showed clearly fishermen reporting declining catch rates, smaller fish, and having to travel greater distances to find fish. Remember, this was all done with baited hooks and dories, not unlike today's recreational fishery. Stocks only increased when war kept fishermen in or very near ports. War was bad for business, but good

David with a bluefin tuna he just brought in. The *Lady Marcia Ann* at the Eastman's dock, Seabrook Harbor, NH, 1976.

for the fish. The fish of
the New England coast by
this time were showing the
effects of the thirty years of
peace after World War II.
I had personally seen the
giant factory ships, always
called Russian by the press,
but made up of ships from
numerous Western and
Eastern European coun-

Trophy tuna tails on the bait shed at Eastman's Fishing Parties, Seabrook Harbor, NH, 1977.

tries as well as Russia, sweep Jeffreys Ledge bare. These ships were not primarily
interested in cod and haddock like American boats, but rather, volume species
like red hake, mackerel, whiting, and herring for reduction to feed the Eurasian
continent still rebuilding from the ravages of the war. Nonetheless, the devastation was real. This paper received considerable written comment. Apparently, I had struck a nerve. I could tell by the comments the readers were not
wholeheartedly buying my thesis, but the research and citations were sound and
extensive, and I was awarded an A for effort. This completed my academic career.

During the winter, the Eastmans had entered negotiations with their River
Street competitors to purchase their business, which had never really recovered
from the bombing. The *Jubilee* had been repaired and certified by the Coast
Guard as fit for service, but their other two vessels had failed hull inspection and
would require massive capital investment to be certified for service. They simply
had run out of money. The Eastmans would get the three boats and the land
and pier on the harbor side of River St. What this meant for me personally was
moving up from the *Lady Lil* to the *Lady Lil II*, which had been Bill's vessel when
I first started working for him. My friend, Erik, was hired to run the *Lady Lil*.

Just before the conclusion of the academic year in 1975, I attended a BU
job fair given by the biology department. Most BU biology students were pre-
med and applying to medical schools, but I was one of the few people interested in research biology. I discussed a job with the research director of the
New England Aquarium working in their research laboratory on a contract for
studying menhaden for the Pilgrim Nuclear Power Station in Plymouth. The

An epic struggle at sea on a stormy, successful day hunting bluefin tuna. Captain David Goethel and Captain Erik Anderson preparing tuna they just landed, ready for sale and transport. Seabrook Harbor at the Eastman's dock, 1977.

contract was for six months beginning in September and the pay was three dollars per hour. If my work was acceptable and the contract extended, we would discuss a permanent position. With my first real job guaranteed, I turned my attention to fishing. I did not attend the BU graduation as no one in my family was interested in going. My diploma arrived in the mail and my school days and academic career were over. The summer passed quickly and uneventfully. I enjoyed running the *Lady Lil II*. My girlfriend visited often. Life was good.

CHAPTER 7

Biology Is Easy

My father had divorced wife number two over the summer of 1975. I was enlisted to move him to an apartment in Newburyport. Why Newburyport, I have no idea, but nothing he did at this point seemed to make much sense. He immediately began beating the bushes for future wife number three. One of the side effects of his new residence was we both rode the bus to Boston. As I remember, the bus cost six dollars per day round trip or thirty dollars for a work week. This was an important consideration for a man making three dollars per hour.

Another new thing for me was paying a car loan. My Torino, it turned out, was made of reclaimed metal. I learned this when I tossed my toolbox in the trunk and it came out with a crash on the ground. I lifted the trunk mat and discovered all the metal was gone and the fuel tank was suspended from a couple of nearly rusted-through supports. Checking under the floor mats in the passenger compartment revealed much the same condition. The left front fender was showing rust spots like cancer as well. Repairs to pass inspection would cost more than the car was worth. So reluctantly, I drove my Torino to the local Ford dealership where I traded it in for a red Ford Mustang II and a car loan.

I also needed a place to live. I had hoped to rent an apartment in Seabrook where there were some huge apartment complexes near Interstate 95 and rents were reasonable, but no units were available. I finally settled for a two-bedroom apartment in Hampton, one town further north, one bedroom more than I needed, and seventy dollars more than I wanted to spend, but it was the only apartment available. I purchased a bed, couch, TV and other necessities. I had enough money left for rent and car payments into early next summer. Some longer-range planning was in order.

I had never worked at anything but fishing and getting used to the aquarium's corporate structure took some time. The research director was the overall boss, but he mainly confined himself to obtaining grants and contracts to keep the lab functioning and to dealing with the other corporate elites in the building.

The lab was overseen by a foreman named Al, who shared a somewhat similar background to mine, having paid his way through Boston University working at various jobs on the Boston waterfront. Most of the eleven employees were chemists who worked on long-term contracts surrounding various contamination issues in Boston Harbor. My immediate superior was a fellow biologist in charge of the grant which employed me. Frank was a very good biologist who had already solved some of the basic issues such as procuring the fish and successfully keeping them in captivity, but he was irreverent to authority and either barely ahead or slightly behind the hot water his attitude produced in the aquarium corporate establishment.

The problem we had been hired to solve was that menhaden, a member of the herring family which feeds on plankton, had a bad habit of swimming into the discharge canal of the Plymouth, Massachusetts, nuclear power plant, where they spontaneously died and then washed up on Plymouth beaches. Menhaden, or as they are locally known, pogies, travel in large schools with many thousands of fish per school. If a school entered the canal, many thousands of fish floated out dead. The problem was compounded by the fact that pogies contain a great deal of oil which causes them to smell most foul during decomposition. The plant had a public-relations nightmare, and they either wanted it solved or a report that said they weren't responsible.

My job at the aquarium, at least at first, was to do water-quality tests on our captive fish as well as feed them to make sure they lived. Our fish were kept in a twenty-foot-diameter swimming pool behind the main building connected by pipe to the aquarium's open system that took water from Boston Harbor. Most of the tests could be run in the research department's laboratory, but one test required specialized equipment housed in the lab one floor below, which checked all the tanks. This lab did dissections of dead aquarium fish and dead marine mammals under contract to the National Marine Fisheries Service (NMFS). The lab was headed by Dr. Sal Testaverde, assisted by Dave Lake. I recognized Sal's name from his family's boat, the *Linda B*, which fished regularly for whiting off Seabrook. He was glad to have a fellow refugee from fishing in the building.

On weekends, I was still carrying charters for Eastmans. By the end of October, my boat had been hauled for the year, but the all-day boat kept running until November 15, and there was one last charter on November 11. Why do I remember the date after all these years? Well, on November 10 a northwest gale swept the Great Lakes with hurricane-force winds, sinking the iron-ore carrier *Edmund Fitzgerald*. This was a 729-foot ship sunk on a lake by 35-foot seas. What do you think would happen to a 58-foot party boat roughly the same distance from shore? I knew I had no interest in finding out.

November 11 started sunny and unseasonably warm with virtually no wind. Lester canceled the all-day boat, but a few people had showed up because of the nice weather and were not happy. On top of this, the charter people arrived and some still wanted to go out despite the weather forecast. The terms were laid out as starkly as possible: we would fish for rock cod within two miles of the beach and the boat would return to port at the first sign of wind, with no refunds. The charter took their refund and handed it out to the participants, but about twenty people still wanted to go out between the two groups. Lester did very little inshore fishing and had no interest in running the trip. He told me to take his boat because at least I had experience on the inshore grounds. I knew at that time I should refuse to go, but I was still an employee and wished to keep my job.

The *Lady Marcia Ann* was a beautiful boat, constructed in Deltaville, Virginia, by the Price boat builders. She was recently repowered with a new 8V71 Detroit diesel and handled like a sports car. We loaded up the people and left. The first stop yielded about six nice cod around ten-to-twelve pounds and a few smaller ones. The second stop had produced about the same when the weather channel on the VHF reported the 9 o'clock weather from Concord, NH, as sustained winds of sixty miles an hour with gusts to seventy-five. This was the nearest station to the northwest of our position. A few minutes later Hampton Beach, roughly two miles to the northwest, erupted with a cloud of sand extending perhaps three hundred feet in the air.

Without even telling the people to reel in their lines, I started the engine and told the mate to start the gas-powered anchor hauler and haul the anchor like his life depended on it because it probably did. Halfway through the anchor retrieval, a wall of sand and wind hit us literally like a freight train. The boat shuddered and I increased the throttle substantially but still could not take the strain off the anchor line. Just when I figured the anchor hauler would pull right out of the deck, the anchor pulled out of the bottom and the mate was able to retrieve the remaining line and the anchor. Even with the engine nearly wide open we were only making about five knots against the wind. I had to close the pilothouse window because the sand was blinding me.

We successfully gained the harbor entrance but one more challenge remained. The tide was low, which meant all boats on moorings took up more scope and swung more wildly on their moorings in the wind. Getting through the mooring field without hitting a boat or sucking a mooring line into the propeller would be a challenge for the ages. The plan was to start upwind against a moored boat's stern, let the wind drift me towards the boat on the lee side, and then use a massive burst of throttle to jet between the two boats and maintain steerage. I did this three times, but each time built up more speed from the bursts of throttle. The final turn was hard to port followed by a downwind landing. The proper way to land a boat is bow into the wind, but this was impossible at Eastman's dock at low tide because there was no water beyond the dock. The wind was almost dead astern. The stern would have to be slightly past the wind so the boat would fall off onto the dock. If the turn was not successful, the wind would force the boat to go broadside onto the sand where no amount of power would remove it. I had to use almost full throttle to complete the turn to port. Once I judged the stern to be past the wind, I went from forward to reverse without stopping in neutral and advanced the throttle, and advanced it some more, until it was nearly wide open. The wind was still propelling the boat forward. The boat's bow could travel about thirty feet past the end of the dock before it would

ground out. I just did not want to hit too hard and damage the hull. The boat had slowed dramatically, but it never stopped until the bow grounded out. The crewman leaped onto the dock and quickly tied the stern tight to the pier before the wind could force it out and we were secure. I had never landed

Underway, David Goethel at the helm of the *Lady Marcia Ann*, on a half-day mackerel trip with Eastman's Fishing Parties, 1978.

a boat under such conditions and have no intention of ever being put in such a position again.

Our work at the aquarium continued to investigate the cause of the fish die-offs. Recently dead fish were dissected, and all appeared to have a common trait, embolisms in their gill filaments. Samples were removed and the gas was identified as nitrogen. The fish were basically getting the bends, the same as scuba divers who ascend to the surface too quickly after a dive. The next step was to determine whether the change in pressure, change in temperature, or both, as the water was pumped through the plant were the cause. The answer appeared to be both, and after consulting with the plant's engineers, it was apparent neither temperature nor pressure could be changed sufficiently to solve the problem. The next question was, could the mouth of the discharge canal be fitted with a small mesh net the menhaden could not swim through? Again, the answer came back no; the water velocity was too high and once algae began to coat the net, the net would alter the fluid dynamics of the canal. The engineers then asked if we could come up with a way to drive the gas out of solution. After a few small-scale experiments at the aquarium, we determined that bubbling compressed air through a diffuser lowered the nitrogen level in the water. Now all we had to do was build a small-scale device and use some of our captive fish to prove it. Easier said than done.

After some discussions with the lab foreman, Al, I was tasked with building the experimental device. Why me? Well, I had already been asked to handle an assortment of odd jobs, like getting the aquarium's antiquated research boat running when their hired captain, who had a penchant for alcohol, was a no-show. During that episode, I demonstrated a fundamental knowledge of wiring, plumbing, and mechanics that all fishermen must possess to keep boats operational. I would construct the device, Frank would assist, and Al would order the supplies. We would use the aquarium's well-appointed machine shop, which was located in the sub-basement under the giant ocean tank.

Al liked me because, unlike most of the people in the lab who complained about having to do anything outside their field of expertise, I would at least try to do whatever needed doing. He still had strong ties to Charlestown, South Boston, and the Quincy waterfront where he had grown up. During this period in Boston's history, these were not areas for the uninvited to wander around. One

Friday night, Al asked me to accompany him to meet some of his friends from his youth. Al had a convertible sports car which he drove with the top down even though the calendar said late November. We started at a bar in Charlestown where he was warmly greeted and I was introduced as "all right," which probably meant safe, not a cop or a snitch. After a couple drinks, we motored back across the Mystic Bridge, down the Central Artery and Southeast Expressway to a marina in Quincy. There we met the owners who lived in a trailer on the premises. Apparently, Al had called ahead because we were treated to a delicious dinner and several hours of conversation about boats (always broke down), boat owners (always wanted everything fixed for free), and boat captains (always destroying the docks with their lack of skills). We left their company around nine and headed into Southie. Again, Al was greeted warmly, and I was vetted as safe. By around 10:30, I was becoming concerned that this soiree could end badly. I told Al there were no more buses and my car was in Newburyport, some fifty miles away. His response was not to worry, he would drive me there. I was not so sure, but off we went, top still down, radio blaring. By the time we arrived, I could feel neither my hands, feet, nor face, but Al was in fine form. He dropped me at my car and roared off towards Interstate 95 for his drive back to his home. As I drove home to Hampton, it dawned on me that the aquarium and the Eastmans were not that different. Lester and Al would get along famously.

CHAPTER 8

Politics Not So Much

Work on building our experimental device, which was basically PVC pipes with holes drilled at precise intervals which would be fitted into an experimental tank, continued for about a week while the experimental protocol was devised and approved. The plan called for using a tank in the cold-water marine gallery on the third floor, which was at the end of the line of the water-delivery system. This accomplished two items, the fish would start at the same water temperature as their own tank, and the water could be discharged through the building drain system to be filtered in the building's filtration system before being returned to Boston Harbor.

Our last issue was moving the fish from their outdoor holding tank to the third floor of the building. We would use large plastic bins, with wheels, filled with seawater which would be wheeled through the public viewing area to a freight elevator for delivery to the third floor. Our access to our fish was through an emergency exit which was alarmed should it be opened. A large console just inside the main entrance contained an assortment of sensors and alarms. This console was staffed primarily with interns on work study who answered customer's questions and monitored the systems. The interns were overseen by a young woman named Diana who seemed to enjoy flirting with my boss. To access our fish, either Diana or one of the interns would have to deactivate the alarm. We would always spend a few minutes talking about our work and everything seemed quite harmless, but some corporate type had observed this a few times and Frank was called in for another lecture about responsibilities in the workplace.

Our initial experiments were to determine the fish's thermal- and nitrogen-tolerance levels. These tests often lasted up to twenty hours. These experiments helped provide me with extra revenue, which with my low base salary proved quite helpful. I was moving fish every day, which meant talking with the girls at the alarm console daily. One girl who had waist-length blonde hair, a perpetual smile, and an insatiable curiosity about the biological aspects of our experiments caught my attention. Her name was Ellen. Frank knew of my newly single status, and, unknown to me or the young lady, had decided to engage in a round of matchmaking. He invited several people from the building to a happy hour at a restaurant named the Winery on the adjacent wharf to the aquarium. Diana brought Ellen and Frank brought me. Frank and I were the last to arrive, and, conveniently, the only free seat for me was next to the stunning intern with the long blonde hair. This was a large group, maybe eight people, so unless you wanted to shout at somebody at the far end, you were confined to talking to your neighbor.

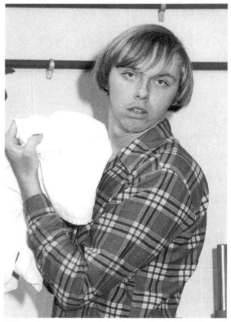

What does a research biologist look like? David Goethel as a research biologist at the New England Aquarium, Boston, winter 1976.

Ellen seemed a little nervous, as she was the youngest person in attendance. I learned she had never had a cocktail, had some food allergies, and lived in the Allston section of Boston. She was nineteen and attended Northeastern University on work study. Someone had ordered a round of drinks and a mixed drink soon appeared for Ellen. Someone else had ordered several pizzas. What could possibly go wrong? She probably drank the drink too quickly and it lowered her inhibitions. She had always wanted to try pizza and bit into a slice. The effect was immediate, and her lips and mouth began to swell. Ellen took a Benadryl to counteract the allergies, but that only enhanced the effects of the cocktail. A while later, the party broke up and people disbursed for the weekend. I walked Ellen to the subway station and asked if she wanted me to ride home with her for safety. I was afraid she would be an easy mark for some of the lowlifes that inhabited these trains. She assured me she was all set, but I asked for her phone number so I could call her when I got home to make sure she was safe. This was not a pick-up line; I was genuinely concerned. An hour later, I called the number and she answered. She was fine and appreciated my concern. After a little small talk, we disconnected, and I sat back to ponder one of the stranger evenings in my life.

The next week, we completed the in-house experiments, which proved the nitrogen could be driven out of solution by bubbling compressed air through the water. The next step was to build a half-scale mockup "bubbler" and test it in the Plymouth discharge canal. I would build the bubbler, which would cover half the diameter of the canal, and a commercial air compressor of the sort used in construction would be delivered to the canal's edge. Measurements would be taken on either side of the bubbler, and from this, calculations could be run to determine how many compressors would be needed to achieve the desired effect. I was back in the basement cutting up PVC pipe. Frank was writing up the experimental results and tending the fish. I asked Frank to tell me when Ellen was working at the desk because I wanted to talk to her. This produced some good-natured ribbing, but knowing Frank, I am sure he went to immediately inquire about the intern's work schedules.

The aquarium had virtually no windows. The whole research lab had one small window in one office. I arrived for work in the dark and left in the dark. One day, the weather had called for partly cloudy and cold when I departed for work. When I left to go home, there was six inches of partly cloudy on the ground. How did I not know this? I began to feel alienated from the natural cycles of the world and was beginning to resent this indoor life. Several days later, Frank told me I should go feed the fish. He had a phone call to take. I knew the fish had already been fed but went to feed them some more. Ellen was working at the desk but was handling customers when I walked up. I reset the alarm and went out to throw the fish a handful of food they did not need. When I returned Ellen was not dealing with anyone. Not knowing how much time would be available, I got right to the point. I had enjoyed meeting her and would like to ask her out for dinner and a movie on Friday or Saturday night. Her face contorted slightly, and I figured I was about to get shot down. Instead, the answer came back that she was busy on Friday night but would enjoy going out on Saturday evening. I asked her to draw a set of directions to her apartment from Commonwealth Avenue, and I would pick her up at seven. Good for me; now all I needed was a place to eat and a movie to see.

The eating part was easy. The Union Oyster House had a restaurant near the movie theaters. They served excellent fresh fish in a variety of ways that would allow someone with food allergies to find a suitable meal. The movie was more complicated, because I never went to movies except an occasional action flick. I had heard girls like romance movies, so I checked out the newspaper and settled on a Middle Ages period piece that had won a bunch of awards for its English scenery. It also helped that the theater had a parking garage and was close to the restaurant. Now, Boston had changed a lot due to urban renewal during the late1960s. The red-light district was in an area known as Sculley Square. This whole area was bulldozed flat and Government Center was built on its rubble. However, being the world's oldest profession, the girls relocated to the theater district along with their relatives, the peep shows and theaters that definitely did not show romance movies. The area had taken on the name the "Combat Zone," and I was concerned with what perceptions Ellen could get from the street theater that occurred day and night. However, the police always had a heavy presence for the benefit of legitimate businesses and their patrons, so we had no issues.

Ellen and I had a leisurely fish dinner and got to know each other better. She was happy that I had had the consideration of finding a restaurant which served fare she could enjoy. We made the movie, named *Barry Lyndon*, just in time for the opening credits. The paper had been right, the scenery and cinematography were stunning, but the plot was a snooze. Ellen had been holding my hand and leaned over to kiss me. We lost interest in the movie until the lights came up. We retrieved the car and retuned to Ellen's apartment, where I was invited in to meet her roommate. I was told to step lightly because the apartment at the end of the hall housed Boris, the local drug dealer, and his two Doberman pinscher guard dogs. I was told Boris did not bother the tenants and the apartments were safe because the local hoodlums did not want to be on Boris's bad side. We sat on the couch, talked quietly, and made out until about 3:00 a.m. I could not believe how much we had in common. I was allowed to sleep on the couch. When we awoke the next morning, Ellen and her roommate made us all breakfast and I left about 10:00 a.m. On the way out, I met Boris and his two companions. He wanted to know who the new guy was who'd spent the night. Apparently, we were not as quiet as we thought.

By early March, all our equipment was in place in the Plymouth power-plant canal. Frank and I drove down from Boston and the people who owned the compressor fired the beast up. The machine had a muffler, but even so, it would wake the dead a half mile away. Some switches and valves were thrown, the machine got even louder and the bubbler I had built came alive, putting out a cloud of compressed air bubbles across half the canal. We let it run for about twenty minutes before taking measurements, which we made in different places on both sides of the device. The goal was twofold: both to see if the saturation levels varied in time and space, and, if so, to have enough measurements to determine a statistical average. Also, several pictures of the device, machine, and measurements were taken for the report. The good news was the measurements confirmed the saturation level was lowered; the bad news was a quick calculation showed you would need about thirty-six of these machines running twenty-four hours per day to correct the issue. We had a solution, albeit in my mind, one with no practicality.

The next several weeks were spent writing and editing our final report. Frank did the writing, and I did some editing as well as taking care of the fish. Boston Edison would be provided a *Reader's Digest* version of the report in advance. Ellen and I were spending almost all our free time together and a good deal of the aquarium's time as well. I had never met anyone with whom I shared so much in common, and we discussed everything imaginable. We were both in love and everyone knew it. But I was pretty sure my career at the aquarium would be over when Boston Edison read our report. The calendar showed the end of March, and the Eastmans were pressing for a commitment for the coming season. On the last Monday in March, I arrived for work and was informed by Frank that the research director was in a meeting with some Boston Edison corporate executives and their engineers. I said, "I take it they have the report, and they are not happy?" Frank said, "I guess we'll find out."

About an hour later, Frank and I were invited into the director's office to join the party. Pleasantries and introductions were exchanged, and the head suit got down to business. Was the report accurate? Yes. Could parts of the report be altered to form a different conclusion? No. Could the conclusion be inconclusive in return for a long-term commitment for you to study the problem further? Now there was a pause in the room. Frank looked stunned and I could not tell if it was because he could not believe what he was hearing or he was trying to form an acceptable response. Before he answered, I spoke up. I knew what I heard and that was a bribe. I stated the results were accurate and their plant was the problem. For my part, I told them, the report will be forwarded to the state. Should I find the report had been altered, I will inform the state of that fact. The head suit looked at me with contempt, informed the research director there was no need of further discussion and marched out of the room with his delegation in tow.

The research director turned to Frank and me, stating that we were in an impossible position. Had we offered to alter the report, the research director would have fired us for lack of scientific integrity and because we had refused, the contract would not be reviewed. The director could assign us to work on learning to run chemical analysis for the chemists or we could look for jobs elsewhere. I asked to have until the end of the day to decide. I wanted to talk with Frank and Ellen.

Frank had surmised this would be the outcome and had already applied for a biology position with the state of Massachusetts. Ellen and I met for lunch, where I relayed the morning's events. This decision would be the first real test of our love. I had no interest in chemistry and was quite content to return to fishing, but how would Ellen feel? She was being asked to commit to a life in a weather-dependent business that she had not even seen yet. Her response was immediate and emphatic. She was committed to me and our relationship. A career on the ocean would be fine. She would help as much as she could, would continue to work at the aquarium and finish her education. I gave my notice that afternoon; my career as a biologist was finished.

CHAPTER 9

Turn! Turn! Turn!

A week later, my career at the aquarium was over. I said goodbye to everyone and headed back to New Hampshire to get the *Lady Lil II* ready for another season. Ellen helped me a lot by doing much of the painting. I still remember her with a big smile daubed with splotches of blue and orange paint. I think she enjoyed this new adventure; her father had taught her freshwater fishing and fly fishing, but she had no experience on the Atlantic. That would soon change. On nights out, we would visit my friends Erik and his girlfriend, Paula, or Marty and his girlfriend, Donna. We also occasionally went to the clubs on Hampton Beach. Here you would find an eclectic mix of Seacoast residents, single and married, out for a night of dancing and gossip.

My father had been busy securing wife number three, who was a temperamental artist. I had been enlisted the previous winter to move him from his apartment in Newburyport to a house overlooking the Merrimack River, and, in keeping with his previous condition, a few months later to a house two doors down. Now he was getting ready to close on a house on Boar's Head on Hampton Beach. I was really getting tired of humping furniture.

Changes were occurring at Eastmans as well. The rod rental concession was turned over to the boats when the man who had handled it previously retired. The captains would buy the equipment and get half the rental fee; the mates would maintain the equipment in return for the other

Ellen dutifully painting the *Lady Lil II* in the Brown's marina, Newburyport, MA, April 1976.

half of the revenue. My crew, especially, was excited by this, as for relatively minimal work, he could expect a big increase in pay. However, on a more somber note, my mentor, Bill, had been diagnosed with early onset Alzheimer's disease. The disease was not debilitating, yet, manifesting mostly at that point just in forgotten names of people he should know, but the doctor told Lil that they should prepare for his retirement from running a boat and consider retirement from the business. This was devastating news for everyone, but at least for this coming season, he was still capable of running a boat, although the doctor suggested more rest and fewer days at sea. From this news, a plan was devised to have Bill run trips only on weekends and only in July and August when passenger loads and revenue were the highest. This would dramatically reduce his number of trips without a significant dent in his revenue. I would take over the half-day walk-on trips in May and June, September, and October.

The *Lady Lil II* was launched in the third week of April, and I was back carrying charters by the last weekend in April. I did not want Ellen accompanying me on charters, as they were almost always all men and having a woman on board could lead to predictable problems. However, the half-day walk-on trade was a different beast. Ellen went often, becoming quite proficient at mackerel fishing and often teaching customers how to fish. Her thousand-watt smile was on permanent display as she marveled at the ocean's wonders. When you live on the ocean, you take many sights for granted. Sunrises, sunsets, schools of fish shoaling, whales breaching, and shark fins passing are just a few of the items that filled her with joy. I had never thought about life on the ocean; I just did it. She had thought about life on the ocean constantly and could not have been happier as she set out to make it happen.

The Northeastern University school year was almost over, and when it ended, Ellen planned to take two weeks off from the aquarium and fly home to see her parents, who lived in a suburb of Buffalo, New York. She would be bringing news to friends and family which we hoped would please them. I had asked Ellen to marry me, and she had accepted. However, we would not announce our engagement until she returned, hopefully with her parents' blessing. The minute her plane left Logan, I was lonely. We had always been together since we got seriously involved and I felt lost in our apartment without her. The two-week period passed like molasses in January, but at least we could talk on the phone every night. I met her parents by phone during this period. Her father, Mort, assembled hydraulic valves, some of which were used in the space program. Ellen's mother, Gloria, ran the school-lunch program at the local elementary school while caring for Ellen's younger sister, Sue Ann, who attended junior high. They were eager to meet their soon-to-be son-in-law and planned to visit during Mort's summer vacation.

I picked up Ellen at Logan from an evening flight. The hug and kiss that ensued when she jumped in the car lasted long enough to fire off the notorious Boston drivers, who laid on their horns and extended middle fingers when our car did not move in what was, to their considered judgement, an appropriate amount of time. I had purchased an engagement ring with a gold band and a natural pearl as a symbol of the ocean. Unfortunately, the ring size was much too big, and we had to send it out to be sized. This would take a couple weeks. We decided to officially announce our engagement at Eastman's big Fourth of July cookout.

I still had an occasional day off in early June, and Ellen and I made arrangements to make a trip to Needham to introduce Ellen to my mother, followed by a stop in Newton to meet my grandmother. Ellen, like all people in this position, was worried one or both might not approve of her. Why, I have no idea, but she was still stressed.

Since graduating college, I had only been home once or twice, so changes would be noticed. To put it bluntly, my mother looked awful: bloated, pale, constantly rubbing her chest. Even a lay person could see serious medical issues, but she resolutely refused to see a doctor. At this time, my mother had a big German shepherd named Brutus, who really liked me. He sensed a change and made a point of inserting his huge head between Ellen and me as we sat on the couch. He just sat there emitting a soft low growl. Ellen petted his head, but being in such close proximity to fur set off an asthma attack and we had to cut the visit short. My mother was very happy to meet Ellen and wished us well, but she stated resolutely she would not be able to attend the wedding. Sad, but predictable. We said our goodbyes and headed over to Newton to visit my grandmother.

My grandmother, Mamie, was in her late seventies and still sharp as a tack. However, she had colon cancer, which she had chosen not to treat, and as a result had lost weight. She was always thin, but now she was gaunt. Mamie took Ellen on a guided tour of the keepsakes in her apartment, giving her detailed instructions on who should be allowed to have certain items. The tour took considerable time, but Ellen was always patient and kind. This was just her way. They finally returned to the living room. Mamie, in her usual direct way, looked at me and declared of Ellen, "You have a keeper, make sure you take good care of her." To Ellen, she stated that I was wicked smart and could be a doctor or lawyer: "He is still young and there is still time." She was clearly never impressed with fishing or biology.

Fourth of July drew near, and I asked my friend Erik to pick up the ring on a day when he did not have a trip. The Eastmans always had a large cookout/pot-luck on the Fourth after the boats got in. All the women were great cooks, and main-course items often included roast turkey, beef, or ham as well as numerous fish dishes. This was a large event, often with more than one hundred people attending. After the majority of people had eaten, we stood together to make our announcement. I placed the ring on Ellen's finger to great applause, hugs, and handshakes. From there, the summer passed quickly with Ellen in near constant contact with her mother over wedding planning. My only job was selecting a best man and ushers and obtaining tuxedos. I got off easy. Erik would be my best man, my brother, Fred, Bill, and another friend, Marty, would be the ushers. I got some grief from my father for not including him, but the fact was, I could not trust him. His alcoholism and other issues just made him too erratic.

As summer moved into fall, thoughts again turned to tuna fishing. My crewman had returned to UNH. Ellen had taken the fall semester off for our wedding, so we made tuna trips together. I taught her the basics of getting the lines in if we hooked a fish and what to do at various points in the process when you needed an extra set of hands. One trip stands out. We made a trip to the eastern side of Jeffreys Ledge, down towards Rockport, about twenty-five miles from Seabrook. There were four or five boats fishing there. One boat hooked and landed a tuna near us, but everybody else kept catching blue sharks, including me. The weather was perfect, one of those fine, sunny, warm September days that signal the end of summer. I was in the wheelhouse, intently studying my fathometer because I had several tunas down very deep, just a few fathoms off the bottom. I was formulating a plan to get their attention when Ellen asked me very calmly to come see something. I went over but saw nothing. She stated very emphatically that there was a very large shark she thought was a mako that had come out from under the boat and eaten the handful of chum she had just tossed in the water. I assumed this was a blue shark. They can be up to twelve feet in length, have an impressive set of teeth, and would probably rattle someone when they were only a couple feet away. We caught them all the time. They did not fight much but often made a tangled mess of the equipment. They had no commercial use; all you could do is pull them to the surface and cut your line off near the hook. A few fishermen carried firearms and shot the sharks, but I never had an interest in killing animals that I had no commercial use for. Killing fish for food was a necessity; killing fish for spite was murder.

I was rigging up a deep-sea rod with a bunch of pieces of chum tied on over twenty feet of line. I would lower this to the bottom. Hopefully the tuna would strike it. The line would break, but maybe it would trigger a feeding response in the tuna, and they would rise to grab a tuna line. At this point, Ellen once again, with more urgency, asked me to come see the shark, but once again it was gone. I told her I was busy, to stop chumming, and hopefully it would go away. Ellen had been sitting straddling the rail with one leg on the outside of the boat. Her foot was only a few feet off the water. Apparently, the shark had grown used to being hand fed, and when no more food was forthcoming, it stuck its head out of the water while snapping its jaws shut. Ellen was scared and angry. She told me to do something about this shark that had just tried to bite her leg off. Blue sharks do not jump, and I now had an uneasy feeling we had a much larger, more dangerous and unwanted friend. Not really having a plan, I grabbed a handful of chum and threw it out far enough from the rail that the beast would have to swim all the way out from the shade of the hull. Sure enough, the shark casually swam out, ate all the chum in one gulp, and returned under the boat. The fish was ten-to-twelve-feet long, two feet or so in diameter, and with a mouth that would hold a twenty-gallon trash barrel. All I could stammer was, "Well, that's a shark!" Because the teeth were conical and not triangular, I deduced the shark was a mako, a definite maneater. Makos have a reputation for great strength and for jumping entirely clear of the water when hooked, often landing in your boat or crushing it. My friend and former captain Dick Muise had fought one for seven hours several weeks earlier, emptied a full cylinder from a .357 into the area forward of its gills, and then fought it for another hour before finally subduing it. Makos are good to eat, tasting like swordfish, but I had no appetite for a fight with one. We hauled in our lines, hauled the anchor, and moved.

On the way home, I decided to stop and fish on a hump named Halfway. Why the name Halfway, well it was halfway between the New Hampshire shore and Jeffreys. I had only fished tuna there once before, having caught four blue sharks in an hour. I told Ellen if we got a blue shark, we would call it a bad day and go home. We set up on the southeast corner where the edge was the sharpest and waited. At least the blue sharks did not taunt us, and after about an hour one lone tuna appeared on the screen. One lone, hungry bluefin. A line soon went off, and Ellen retrieved all the other lines while I fought the fish. When I got winded, I would hand her the line and ask her to just try to hold on. When we got the fish to the surface, I did the same thing, had her hold the swivel while I darted the fish. She then provided me a tail strap and the fish was ours. Our arrival in Seabrook produced a mild stir as word quickly got around that I had a girl crew, but the truth is Lil Eastman had done the same with her young husband thirty years earlier.

Our wedding was in mid-November. We left Hampton the week before the day of the wedding for the ten-hour drive out the Massachusetts turnpike and the New York state throughway. Ellen did not drive; her father thought women did not need licenses or cars, so I drove the entire way. I quickly named the New York part of the drive the world's most boring road as there was nothing to look at and the exits were twenty to thirty miles apart. We arrived around 8:30 and as quickly as possible I went to bed, exhausted from the drive, the last sixty miles having been in moderate snow. My impression of wedding planning was roughly what generals must do before launching a military offensive. Everything is planned right down to the smallest detail, and then you just hope nothing major goes wrong. We had, what was considered at the time, a mildly nontraditional wedding. In addition to readings from the Bible, we played music, most notably the Byrd's "Turn! Turn! Turn!" which is verses from Ecclesiastes 3.1–8 set to music. This description of life had always resonated with me and still does. A reception with a meal and dancing followed. Everything went off without a hitch, something that, as I was to learn from future weddings I attended, was not always the case. We arrived at our hotel room after midnight, both exhausted and elated. We were husband and wife. Ellen and I would fly out early the next morning to Seattle to begin our honeymoon. We tried to visit as many national parks as possible because we both figured we would never be here again. We had a wonderful time seeing things that may not interest most Americans but that caught the attention of aspiring biologists, with a few stops to check on West Coast fisheries and harbors.

CHAPTER 10

Never a Dull Moment

We returned to New Hampshire in early December. Ellen started work at the aquarium and was enrolled to start her junior year at Northeastern by completing the winter and spring semester. My friend Marty offered me a job working for his moving company in Portsmouth. One of the perennial drawbacks to the party boat business was its seasonality. You had a good-paying job for six months but always needed winter work. Some people, like Bill, relocated to Florida, where he worked as a welder for a citrus farm. Others found work on boats in Florida. Some were teachers, others plowed snow, whatever you could find. Moving furniture was not glamorous; in fact, it was plain hard work, but it helped pay the bills. Ellen continued her schooling, and I went off to work every day.

Ellen soon grew tired of apartment living. As she rightfully pointed out, we could own a home and pay the same amount of money on a mortgage that we paid in rent and not have to listen to daily disputes through the walls. I was not sure we had enough for a down payment, and even if we did, our cash reserves would be nonexistent. She asked if she could find a realtor and at least talk about what we could afford. We both agreed we wanted a Cape Cod-style house. We also agreed we wanted to stay in Hampton because the schools were good, and someday we wanted kids. We spent early 1977 house hunting. We narrowed the search down to two Cape-style houses with about the same asking price. Ellen was a better negotiator then me and bluntly told the owners the price would have to be about 10 percent lower for what the bank would loan us. One couple balked, but the other people were paying mortgages on two houses and had to shed one or face financial consequences. They met Ellen's request, and on a cold weekend in late January, I finally moved our furniture into our own house.

I had also been discussing buying a commercial gillnet boat with a captain of a party boat in Hampton. Rocky Gauron and I had looked at a forty-two-foot gillnetter to become partners in as we both needed winter jobs, but I had to back out because all my money was now in the house. Even though we were competitors on the water in the summer, the fishing community is small and always worked together. Rocky took the news well and continued to search for a boat within his price range over the coming months. Spring rolled around and the party boats were put back in the water. One day towards the end of May, a weather situation like November of 1975 developed. The weather forecast called for northwest winds of fifteen-to-twenty-five knots while I was out with a charter. We used to joke, with gallows humor, that you should add the two numbers together to get what would truly happen. Today would prove no exception. My anchor broke loose out of the bottom, so I decided to idle towards shore some fifteen miles to the northwest and find some hard ledge where the anchor would hold. However, at an idle, the boat was not making headway. When I advanced the throttle, the engine quit. I quickly took the primary fuel filter apart, found it empty of fuel, refilled it, and started the engine. The same thing happened again; small amounts of fuel would flow through the line but calling for additional fuel only burned what was left in the filter.

I was in a jam; we were drifting further out to sea and the sea was building rapidly. I called Brad on the nearby *Lady Merrilee Ann* (ex-*Jubilee*) to request a tow. I figured if he could tow directly into the northwest chop, I could work on the fuel system to troubleshoot the problem. I had a good crewman named Gary and we soon had the towline secured and the boat headed into the wind with very slow forward speed. I tried everything, but the engine would not run. Our situation was still deteriorating as the wind continued to increase. I relieved Gary of the wheel, told him to try and get the people spread out to make the ride better and began talking to Brad and other captains on the radio. Brad had tried to tow me towards Seabrook, but that course caused the waves to hit the starboard quarter. The boat pounded and shuddered, and we both agreed that could be catastrophic. So, we decided to proceed straight into the wind. Brad's boat was bigger and would break the oncoming sea some. Our course would bring us near the Isles of Shoals, which would act as a lee and once we exited that area, we would have about four more miles of misery before we were close enough to the mainland to have a permanent lee. Then we could safely turn broadside to the wind and finish the tow to Seabrook. Even though we were incrementally getting closer to shore, the wind kept increasing and the sea continued to grow.

Sinking, it is often said, does not usually occur from one catastrophic failure, but rather a series of incremental failures in rapid succession. The boat was more like a submarine, as it was covered with water from bow to stern. I told Gary to continually check the bilge pumps to make sure they were working. He reported one was pumping continually and another was spitting water. A third one would pump for a minute when he turned it on but run out of water. We were taking in water some place, probably through deck seams, but it was not yet catastrophic. About twenty minutes later, Gary was back with a new problem. A member of the charter claimed to be having a heart attack. He said the rest were getting extremely nervous, some near panic. Even Gary had a look of fear. I looked at him with a serious expression and said, "Well, if he dies throw the body overboard, we need the stability." His eyes went wide. "Are you serious?" "No," I laughed, "just put him on the engine box and cover him with some coats. And tell the rest of them to spread out. All this weight in the forward third of the boat is keeping the bow from going over this sea." Gary's face brightened for a minute, and he said, "I thought you were serious!"

The tow continued for two more hours before we finally got close to the Shoals. There would be about a twenty-minute period where it would be nearly calm. This would give the nervous and the seasick on both boats a chance to recover before being subjected to a further beating as we exited the lee of the islands. The worst was over, but it would take another three hours before we secured the lines in Seabrook. Lil had called Ellen to tell her to get some dry clothes together as there had been an "incident" and I was being towed into the harbor. I assume someone picked her up because she was on the dock waiting when we finally tied up. Gary and I were completely soaked, but the guy with the heart attack made a miraculous recovery when his feet touched the dock. After the customers had departed, I congratulated and thanked Brad for his seamanship and congratulated Gary for handling a very difficult situation professionally. I gave Ellen a hug, took the dry clothes and said I really ought to kiss the ground, but instead changed and went to work with several captains to try to fix the problem. No one could get the engine to stay running either, so at least I had not overlooked something stupid. Our mechanic was called, and the following day he located the problem. A ninety-degree fitting going into the primary filter was plugged. Imagine our surprise when we found approximately two feet of eight track tape jammed in the fitting! How this made it through the filter on the truck, the filter on the fuel tank from which we pump fuel into the boat, was anyone's guess. What we do know is that thirty-seven people could have lost their lives because of the carelessness and stupidity of someone in the fuel distribution chain.

The rest of the year passed uneventfully, and as fall rolled around, we did our customary tuna fishing. On returning from a successful trip to Massachusetts Bay, I spotted a new boat in the harbor. Rocky had purchased a thirty-five-foot gillnet boat, which he named *Denise Ann* after his wife. I had a job for the winter as crew.

Losses and Gains

Rocky had bought a smaller version of the gillnetter we had considered purchasing. The boat had a twelve-inch gillnet hauler and a very basic hydraulic and electronic package. The boat would be suitable for fishing the inshore waters where the tide was not strong but incapable of fishing on the offshore ledges. Gillnets are monofilament twine hung between a float line, which has flotation every six feet, and a lead line, which as the name implies has a lead core. Thus, the floats hold the net up vertically from the bottom in the water column. The nets are usually fifty fathoms (three hundred feet) in length. You tie nets together to get a string of appropriate length for the bottom you are fishing. The net is set over the vessel's stern and sinks to the bottom, where it remains until the vessel retrieves it. The hauler is a circular device which rotates, containing teeth which grab the lead and float line. The teeth are driven by a cam which brings them down on the net as it enters the hauler and releases the net as it goes around. The bigger the diameter hauler, the more teeth grip the net and the stronger the pull. Most boats had thirty-six-inch-diameter haulers which could be used in any depth and hauled the nets at a quicker pace then the twenty-four-inch and twelve-inch models. The nets are hauled on larger boats from a hauling station approximately one-third of the way down the rail from the bow, guided by a roller that extends out from the rail. The net and fish pass over the roller and around the hauler to a table where the fish are picked from the net. The net collects in a pile at the end of the table. After the string is hauled, the net is flaked over a bar to remove twists and tangles so the net will stand up when reset on the bottom. Gillnetting has little bycatch and is very effective at catching larger fish. Our targets were big cod in the spring and summer and large pollack in the fall and early winter.

Most gillnetting was done within the confines of the territorial sea, which extends twelve miles from shore. This is the area the United States government claimed as its own and in which foreign factory ships were prohibited. This was good, because gillnets set in the path of these giant mobile gear vessels were towed away and cut loose when the ship hauled back. Since the early 1970s, some intrepid netters had tried to fish Jeffreys Ledge, which is mostly outside the territorial sea, with decidedly mixed results. I watched a clash one day between a gillnet boat and the Soviet fleet while running a charter. The gillnet captain put his vessel next to one of his marker flags, climbed on the roof of his boat with a shotgun and blasted away at the lead vessel's pilothouse windows. I doubt he even scratched the glass, as the pilothouse was probably five stories off sea level, and the gillnetter's vessel looked like a toy next to this giant steel behemoth, but the encounter encapsulated the frustration American fishermen felt in dealing with foreign vessels. The gillnetter flags went down one by one as the big dragger towed away all his gear. As the vessel passed, there were two men in cook's aprons smoking on the fantail. They were waving, obviously unaware of what was happening.

By the mid-1970s, the situation had become so serious that the commercial fleet, sailing from various East Coast ports, converged on the Potomac River to stage a protest for Congress to pass a two-hundred-mile limit. Passage was no means guaranteed, as the navy and State Department both vigorously objected. The navy feared other countries would follow the lead, hindering free passage of warships and the State Department had entered numerous deals trading fishing rights for military bases since the end of World War II. Nevertheless, shepherded by Warren Magnuson of Washington State and Ted Stevens of Alaska, the Magnuson Fishery Conservation and Management Act of 1976, (renamed the Magnuson-Stevens Fishery Conservation and Management Act when amended in 1996; hereafter to be called the Magnuson Act or just "Magnuson," as it became known in the industry) was passed by Congress and signed into law by President Ford on April 13, 1976. Many fishermen argued this was a Faustian bargain. The hated foreign pillagers were removed, but in return, you were now subject to American bureaucracy. Many feared the words, I'm from the government and I'm here to help you. (A few years later, the two-hundred-mile "exclusive economic zone" [EEZ] became the world-wide standard with the passage of the 1982 United Nations Convention on the Law of the Sea.)

As of mid-October 1977, there had not yet been any discernable changes in fishing. Rocky had to obtain a permit from the National Marine Fisheries Service (NMFS) in Gloucester, but that was about all. There was a minimum mesh size, but we used mesh well above the minimum to catch only large fish. Such was the situation when the *Denise Ann* set sail around noon in mid-October into an increasing southerly wind to set her nets and begin her commercial career. The weather worsened rapidly and by the time we were four miles offshore, our plans changed to just get the gear out of the boat on the nearest set of ledges and beat a hasty retreat.

The three-seven net strings ended up on two large pieces of ledge, the North Grounds and the Speckled Apron, about five miles south southeast of Hampton. People often ask where the colorful names came from, and the answer for the most part is, no one knows. Some have been around for hundreds of years, some were named for fishermen who fished the area extensively, and some were named for how or when they were discovered. The names were simply a way for fishermen to communicate information among themselves, almost like a special language that only they understood. At any rate, the gear was set and would not be hauled for two days because of the weather.

Part of the magic of fixed gear fisheries to me was peering over the side into the depths to see what would emerge. The fish would first appear as ghostly white blobs but grow rapidly in size and shape, as well as color, as they neared the surface. Sure enough, as the net grew near, white images appeared. They were too large to be spiny dogfish, our chief fear, the time of year seemed too early for pollack, and the shape was too streamlined for cod. What were they? Well, as the first fish exited the water, the answer was clear: bluefish, and lots of them. This was a contingency we had never considered. Writhing, snapping, finger-removing bluefish! Someone produced a hammer, and a plan of attack was formed. Rocky was running the hauler and would give each fish a stout rap on the head as they went around the hauler. The other crewman and I would then remove them from the net and place the fish in holding pens. Those held about two thousand pounds, and we had more than that after hauling two strings. What to do? We took all the dead ones out of the pens and threw them under our feet and placed the live ones from the last string in the pens. We were literally up to our knees in fish. After we finished hauling, the bluefish still had to be cleaned. When bluefish die, they turn stiff as logs. They are tough to get a grip on and tough to gut. By the time we were done, none of us could move or feel our fingers. We expected to get little money for the fish, but because of the storm and corresponding lack of fish on the market, we received a good price. Our first day had been a smashing success.

Rocky had kids and commitments, so I was captain on a few trips. We tried to target cod because two hundred pounds of cod netted the boat the same amount of money as four thousand pounds of pollack. The pollack, however, had other ideas, and more than once the net floated on the surface as the pollack swim bladders acted like floats and you could see a ribbon of fish floating into the distance. By early January, the pollack had begun migrating back to the east and the *Denise Ann* was fishing a spot called Two Box Ridge. Why Two Box Ridge? Well, it seemed no matter how many nets you set in the area, you caught two boxes or two hundred pounds of fish. This piece of bottom was only about three miles from shore, and when the winter storms passed through you caught all manner of junk that rolled around on the bottom. Mostly it was kelp and old heads from lobster traps, but, after one storm, our nets were full of cut-off steel lobster traps. Traps at this time were almost exclusively made of wood, but some enterprising companies were trying to get local fishermen to try their steel traps. A local fisherman had bought or been given some, decided he was not impressed, and cut them all off in the vicinity of our nets. Every time there was a storm, the traps would bounce around on the bottom, ending up in our nets. They had started to rust and fall apart and removing them from the net was like wrestling a roll of barbed wire.

One day in late January, it seemed like we caught a trap for every cod. We were tired and discouraged when we returned to port and tied up at Eastman's dock. We were allowed to use the dock in the winter because their boats were all on land. I was surprised when Marcia Eastman came down the dock as we tied up and told me to stop in the house before I went home. Marcia started by saying there was no easy way to deliver the news she had received. My father had called and told her my mother had been found dead in her bed in a wellness check by the Needham police. The cause of death was a massive heart attack. I was stunned and speechless.

I had always expected this would occur, as she had obvious mental and physical issues, but I guess you always expected there would be a period of decline in which you could say goodbye. Here there was nothing, just a fait accompli. Marcia suggested I call my father and I thanked her for having to be the bearer of bad news. I telephoned my father when I got home. He had assumed full command of the situation without asking for any input from me or my brother. I was informed I would not want a viewing as she had been dead several days. Being a lawyer, he had somehow blocked an autopsy, which is required in Massachusetts for an unattended death, and had her body transferred to a funeral home for cremation. He had written an obituary which announced a memorial service in several days. After that, he planned to spread her ashes in the backyard. I was pretty sure that was illegal, too, but I was simply overwhelmed. Why the hurry? In retrospect, my father was probably deep in debt and my mother's death offered him a lifeline. The sooner he had a death certificate, the sooner the alimony checks ended. His offer to be executor of her estate would net him a decent-sized cash settlement from the sale of the house and the furniture. In fairness to him, I think he really did believe he was acting in all our best interests; apparently it never occurred to him that grieving was necessary as well.

I needed Ellen but had no way to contact her until she got home from school. I called Rocky and told him to retrieve the nets or run the boat for the next week while I attempted to sort everything out. When I told her, Ellen was as stunned as me. She immediately called her parents to ask for guidance, but they really had no way to help or knowledge of what to do. Ellen and I moved like zombies, showing up at appointed times, performing required rituals and signing whatever documents were thrust in front of us. Other than a few pictures and heirlooms we were able to bring to our house, within a month it was like my mother had never existed. The house was sold to the neighboring college that abutted the property. They would move the house to a newly constructed foundation further back from the road and provide extensive renovations in keeping with the house's historic significance, which were badly needed. In an ironic twist, the foundation was dug in the area where my father scattered my mother's ashes. She is literally buried under the house she loved so much.

While this was occurring, there were major changes in the works at Eastman's as well. Bill would retire, and Brad and I would split his share of the business, making fixed yearly payments for the next ten years. I was now a minority owner. Also, we formed a new corporation, Seabrook Fisheries, to finance the construction of a new sixty-five-foot vessel for the all-day trips. I owned a 25 percent share of this business. The Eastmans had never built a new boat, but several new, fast, all-day boats had shown up in neighboring ports and customers began to demand more comfort and faster rides to the grounds. We hired a boatyard in Kennebunkport, Maine, run by a naval architect/boatbuilder who said he could build us a sixty-five-foot vessel that could cruise at twenty knots. This was almost double the speed of our current all-day boat. Construction began that winter.

I returned to running the gillnetter the last week of January. We were still fishing on Two Box Ridge and a couple other nearby humps, catching three hundred to five hundred pounds of cod per day. We were leaving relatively late in the day, around 8:00 a.m., because of the short ride to the grounds and the cold weather. I had a chance to watch the weather on Monday morning. Competition among the Boston television stations was fierce, and weather was a major factor for people in deciding which station to watch. Don Kent was the oldest, most seasoned forecaster. He had argued for a weather buoy for years to be placed off Cape Hatteras, as most of our big storms and hurricanes either formed or passed through that region. Bob Copeland was next in age and experience, and Harvey Leonard was the relative newcomer. All made their own forecasts, not relying on the weather service, whose forecasts were parroted back by television weather people in most other parts of the country. They all explained their reasoning well with charts and graphs if people took the time to listen.

This morning they were all excited and intense. Warm air moving up from the Gulf of Mexico would collide with cold air barreling towards North Carolina from Canada. The whole mess would be blocked from moving at its usual speed by a gigantic Arctic high over Labrador, Canada. The storm would move slowly, dropping historic amounts of snow with hurricane-force winds. Coastal flooding would be extreme, both because of the high astronomical tides, but also because of the multi-day hurricane's easterly winds. Harvey Leonard was particularly intense, describing upcoming events as a storm like no other that living New Englanders had ever experienced. Well, they certainly had my attention! I made the unilateral decision to haul our nets and keep them on board the *Denise Ann*. We would not be able to haul for at least four days, the fish would spoil, and the nets would be full of junk. We got in around 2:00 p.m. and by then the wind was about twenty-five knots with intermittent snow. We secured the boat with extra lines and headed home.

We had extra duty that night. The newly formed New England Fishery Management Council (NEFMC) was holding a public hearing for Amendment 1 to its groundfish plan under the auspices of the Magnuson Act. We were distinctly not impressed, as minimum size limits for cod and haddock would be imposed on all fishermen. Imagine having to tell some guy who had "Death before dishonor" tattooed on his forehead that he could not keep the twelve-inch cod he just caught. At best you would probably get punched in the face, at worst? I liked my teeth firmly attached to my jaw, thank you very much.

Fellow captain Brad offered to drive his big Buick station wagon to Gloucester. The thing was the size of a tank and was assumed to be able to handle the snow. By now it was snowing moderately, and the wind had picked up some more, but the plows were still staying ahead of the snow. When we arrived in Gloucester, we were greeted by a whole different world. Ellen opened her door, which was immediately wrenched from her hand, and she went sliding across the parking lot like a kite in a maelstrom. She grabbed a piling in a last-ditch effort to avoid an icy bath in the harbor, and her feet went out from under her. She looked like a flag on a pole when Brad and I caught up to her and yarned her in. When we opened the door to the restaurant, we were greeted by a burst of escaping hot air and the smell of stale booze. The Gloucester fleet had all come in for the storm and apparently had spent the afternoon at the bar. The hearing started and almost immediately ended. An announcement was made that the hearing was canceled due to the weather; those from out of town should find a hotel, those who lived locally should go directly home.

We were screwed. Gloucester had no hotels, and if we were going to drive some distance to find one, we may as well try to go home. As we crossed the open ground in Essex and Ipswich, we began to doubt our decision. The snow was horizontal, and drifts were over ten feet and climbing, but the center of the road was blown clear. We lumbered on, the big Buick occasionally plowing through a snow drift. When we finally reached Hampton, there was probably a foot of snow on the ground if you could measure it. The town had surrendered the side streets like the one Ellen and I lived on. The Buick plowed down the road and was unable to turn into our driveway, skidding to a halt in our front lawn. We all manned snow shovels and finally got the car turned around. Brad headed off to his home across town. He would be our last human contact for four days.

When we awoke the next day, the house was as dark as Grant's tomb. The snow had drifted over the first-floor windows and the rest were covered with snow. The temperature was about ten and snow was still falling hard. Amazingly, we still had electricity, but the phones were out. Snow continued to fall all day, all night, and part of the next day. You could not force a door open; escape would have to be out the garage door. The wind blew at tremendous force, shaking the house, but because the temperature was so low, snow was not sticking to the trees or wires. I wondered about the boat and the people who lived near the beach, but with no phone, you could learn nothing. At this time, we had a rooftop TV antenna. It had either blown away or something, as all we had was static. The radio was not much better, as apparently most stations had been knocked off the air. Finally, on the fourth day, the weather abated, and we could begin to dig out. Not that it mattered much, because the road was snowbound. Ellen and I shoveled and shoveled and shoveled some more. We could not see our car, but knew it was there, completely covered in snow like it did not exist. One full day of work got our car and one door to the house shoveled out. We still had forty feet of driveway to go.

We began shoveling the next morning, and around noon two town vehicles appeared. One was an enormous bucket loader and the other a road grader. The loader would angle through the snow, dumping it in piles on each side of the road. The piles formed walls between fifteen and twenty feet in height. When this was completed, the road grader scraped the road nearly bare until it created a pile of snow which the other machine then removed. I was dismayed; it would take days to shovel the end of the driveway. I recognized the man running the grader as a fellow fisherman. I asked if his friend could take a couple scoops out of the end of our driveway. No one else on the street was out, so they quickly took out four scoops. The mounds on either side of our driveway were over twenty feet high, but by the end of the afternoon we could get our car out.

Phone service also returned. I called Eastman's and Marcia answered. The boat was still there, but very low in the water. The beach and the roads were still impassable, so she advised me not to bother trying to get to the beach. At least the boat was still afloat and the people I knew were all surviving. The Blizzard of '78 had come and gone; we had survived the storm like no other.

CHAPTER 11

Large Fish

Restrictions were finally lifted for local travel on the beach four days after the storm ended. Route 1A from the Hampton border to Portsmouth was still closed because many of the seawalls had washed away along with the roads behind them. The part of the seawall north of Boar's Head that was just sheet piling was destroyed, with giant pieces of steel bent and twisted at macabre angles. The road was heavily damaged but passable. Rocks, boulders, and other debris had been pushed aside. South of Boar's Head, the railing atop the seawall had the same twisted look as the steel on the north side. One could also see that the water had crossed the road, the only time that occurred in my life. All that saved Hampton Beach was the snow drifts, which kept the water from destroying the buildings. However, when the tide receded, this slushy mix froze into glacial-like material. Very heavy equipment had been used to clear a single, rutted path north on Ocean Boulevard and south on Ashworth Avenue. The piles of material, snow, ice, sand, seaweed, logs, rocks, and pretty much anything else you could think of stood twenty feet high as silent testament to the storm's strength. The storm had sustained recorded winds of 110 miles an hour near us and developed a well-formed eye characteristic of hurricanes, though this was an extratropical cyclone.

The *Denise Ann* looked like an igloo, completely entombed in ice. The boat was low in the water because the pile of nets had been continually covered with spray and snow that had basically turned the rear deck into two feet of solid ice. We needed about two hours to thaw the engine and saltwater washdown system and start the motor. Everyone knows ice forms at thirty-two degrees, but saltwater freezes at much lower temperatures depending on the salt content. We could use the washdown system, which cooled the engine by means of a heat exchanger to melt the ice in the boat, starting at the scuppers, which are ports cut in the hull to let water out. Our hope was to melt most of the nets, then go set them, and let the ocean melt the rest. Good in theory, bad in practice. The top net set normally for about one hundred feet before coming to the frozen mass.

The strain of the set part of the net pulled the frozen mass over the stern in a blob along with all the buoys, which, luckily, were still tied to their respective net ends. All we could do was let physics work and come back in the morning to haul the nets, which would thaw overnight. We would have a mess, but at least we still had gear.

Every other netter I talked to, as well as lobster fishermen, lost everything. The gear had dragged some distance, become entangled with other gear, and stopped. It was all gone, like some invisible hand had swept the ocean bare. The losses were catastrophic, and many businesses would never recover. The effects on ocean life were equally severe. Reports began to come in from east-facing coastlines reporting thousands of dead lobsters and crabs were washing ashore, apparently pummeled to death by the waves. I know we caught nothing until early March, when cod, driven by their primal urge to spawn, began to appear south of the Isles of Shoals.

Life on land, however, resumed its normal pace after several weeks. My wife had transferred to the University of New Hampshire to complete college. She was now a resident, and resident tuition was a fraction of the cost of tuition in Boston. She left her job at the aquarium and began teaching figure skating at a rink in Dover, New Hampshire. She was a competitive figure skater in western New York and could skate circles around all my hockey-playing friends while doing jumps and spins. Going skating with her was a humbling experience.

What she could not do in the fall of 1977, however, was drive a car. This had to change, as New Hampshire is a rural area without bus or train routes. Everyone agreed I was NOT the man for the job, so several friends who were her age volunteered for driver education. One, a young man named Richard, seemed to relish the challenge. Richard was unflappable and unfazed by anything. He owned a giant Ford LTD, which also seemed indestructible. The car did have one unusual feature. The horn, it seems, was a strip built into the inside rim of the steering wheel instead of being in the center like most cars. If you gripped the wheel too firmly you set off the horn. One afternoon, Richard instructed Ellen to park in front of a Chinese restaurant on Hampton Beach. His plan was to go procure an alcoholic refreshment to steady his nerves. The parking was head in and there was no curb. The car was enormous, and Ellen had a hard time seeing where the car ended. As she pulled into the space between two parked cars, she held the wheel with an increasingly tight grip. Suddenly, the horn went off, her foot momentarily slid off the brake because she was startled by the horn, and the car lunged forward. She regained control without hitting the building, but the patrons all dove for cover, fearing the worst. Richard shut off the car, calmly walked into the restaurant, and ordered a double rum and coke like nothing had happened. He did, however, on reflection, decide to drive Ellen to our house. Lessons were suspended for that day. Despite the rough start, Ellen passed the driver's test on her first try.

When the fish began to show up in March, they seemed to be all mixed together in size and species instead of having usual spawning runs of fish the same size in each area. This meant the nets were particularly exciting to haul as you never knew what was coming up next. One cold day in March, I was peering over the side as something particularly large appeared deep in the gloom. Was it a shark, maybe a sturgeon or some other marine oddity? When it finally made its way to the surface, the fish turned out to be an enormous cod, the biggest anyone on the boat had ever seen. In fact, its head was so large it would not fit between the bars on the roller. I put a gaff in its head so we would not lose it if it ripped out of the twine, backed the hauler down, and handed the gaff to a crewman behind the end of the cabin. He could not roll it over the side by himself, so I went back to assist. The fish was huge, with the head being more than twelve inches in diameter and the body over six feet long. We rolled the fish under the table and finished hauling.

During this period, we placed the boat on a mooring at the end of the day, loaded our catch into a large skiff which we ran into shore to meet a truck that housed a scale and ice. Usually, the truck was staffed by two people, but that day, only one man was onboard. This man the boat crews had affectionately named Fat Wayne. Wayne was a true giant, over six foot six and well north of three hundred pounds. He ate perpetually and usually had a large submarine sandwich on top of the scale in case he felt a need for feed. We pitched fish into wooden boxes on the back of the truck. The boxes weighed about fifteen pounds and held one hundred twenty-five pounds of fish plus ice. Wayne would take a full box, place the box on the scale, and shift fish around between the boxes we were filling until the box weighed more than the required weight. You could not sell someone a box of fish that weighed less than what the buyer was being charged for, so the boxes always weighed more, sometimes two or three pounds over, sometimes more; that was just the way it was.

Wayne would then move the full box, now weighing one hundred fifty pounds or more, to the front of the truck, where they were stacked five boxes across and four boxes high. The width of the boxes was designed so they would fit very tightly across the truck so the load could not shift while driving. The placement of the last box in the top row was very difficult when working alone. You had to lift the box and slam it into the space at an angle and then put a shoulder into the box to force it in. Once in, the entire row was locked in place. This day I was waiting on the shore with my crew while Wayne finished the boat ahead of me. We heard a blood curdling scream followed by a weak "help me" from inside the truck. Three of us climbed up in the truck to be greeted by a macabre sight at which we could not decide whether to laugh or cry. It seems Wayne had stood on his toes to slam the last box in but did not move back quickly enough. The box had come down on one of his enormous rolls of blubber which was now pinched under the box. He was standing on his toes so his enormous mass would not pull down on the blubber and add to his considerable misery. We decided one man on each side would try to push the adjoining boxes towards the edge of the truck, and the third man and Wayne would attempt to push up the box pinning him in place. On the third try, we were successful, and Wayne collapsed on the floor of the truck. We were now greeted by the day's second strange sight, a giant roll of fat the most hideous shades of red and purple imaginable.

We revived Wayne with his submarine sandwich and a drink, and he was soon back to weighing fish. We did put a crewman in the truck to help move the boxes. Now, on a good day Wayne had no sense of humor. I do not know if he just took everything literally or if he just did not get punchlines. I wanted our giant fish weighed and tried to warm Wayne up with an offer. I would give him the fish to eat if he would weigh it on the scale. I told him I doubted he could eat this fish in a meal. He turned, looked at me, and responded without a hint of humor that he once ate a thirty-pound turkey in a meal. What could you say? The man had raised gluttony to an art form! All I could stammer was that his statement did not surprise me at all. He weighed the fish at 137.5 pounds. When you added back the approximate weight of the guts and gills, we had removed, the whole weight was about 164 pounds, by far the largest cod I have ever seen! This day had certainly been memorable for more than one reason.

Meanwhile, our new party boat was taking shape nicely in Kennebunkport. Building a boat is an enormous undertaking, and one misstep in the order of placing items into the vessel can cause delays and cost overruns. By mid-June, the hull was nearly completed, the engines were mounted along with the fuel tank, and work was underway on the highly technical boring through the hull for the propeller shafts. The boatshed was on a riverbank; a road into the boat-yard ran right next to it, and then there was a very steep hill, almost a cliff, that rose probably seventy-five feet before the ground leveled off. On that level ground sat a house, probably built in the early 1800s, with a second-story, wrap-around porch and a widow's walk on the third-story roof. On the lawn between the boatyard and the house was a giant tree about 150 feet tall and perhaps three feet in diameter. Apparently, the homeowner had decided to have the tree removed. The tree had several large limbs that towered over the house and several that grew on the side of the boatshed. Anyone who knew anything about tree cutting knew whichever set of limbs weighed more was the direction the tree would fall. The people hired to do the felling, however, apparently knew nothing about the job at hand because they planned to drop the tree between the boatshed and house. A chainsaw began grinding into the tree and work on the boat stopped. Everyone came outside to stand a safe distance from the boatshed in case the outcome was not in our favor.

One of the boatyard crew offered up to the tree cutters that they were mak-ing a huge mistake and the call came back to ***k off. In the true Maine way, sound advice had been rebuffed and no more would be forthcoming. A case of beer was produced, and everyone sat down to watch the show. The tree was notched at right angles to the direction it would fall, so when it decided to go, there would be a huge rending of broken wood as the tree shuddered, kicked off the stump, and began to fall. Sure enough, there was a huge crunching crack, the tree wavered and fell with a noise that was probably heard a half mile away. When the dust settled, the widow's walk and third story of the house were gone, crushed into the second story, and the wrap-around porch on the boatshed side was gone as well. The tree was basically in the center of the house at roughly a thirty-degree angle. The boatyard worker who had offered advice let out a raucous well done followed by a huge Bronx cheer. At least the boat was safe. When we came back the following week the tree was gone, and the remainder of the house was covered with blue tarps.

The boat was launched in mid-August, and everyone was excited to go for a ride. When we cleared the river, the throttles were advanced and to everyone's astonishment except the builders, the boat took off like a rocket. The speed at the engine manufacturer's suggested throttle was just a shade under twenty knots. Several days later, after Coast Guard hull and safety inspections, the boat, which was now named *Lady Erica*, would be run down to Hampton Harbor to begin service. This would necessitate captains changing vessels, and I assumed command of the former all-day boat, *Lady Marcia Ann*. Meanwhile, Ellen had graduated from UNH and taken a job as an entry-level biologist in Portsmouth, New Hampshire.

While 1978 was proving to be a whirlwind year for Ellen and me, the year was disastrous for my father. I almost banned him from the boat after an incident in which he and his wife, having had way too much to drink, caused a ruckus. He was jigging for cod and, as he bounced the jig up and down, his

Eastman's Fishing Parties fleet, 1977. Photo by Fred Goethel.

pants retreated further and further, finally falling to his ankles. You would think his wife would have been embarrassed and made him pull up his pants, but she seemed to egg him on. I did not see this but, a man with kids did, and he came to me to complain. I told him I would take care of the problem and I did, hissing at both in cold fury to make sure his pants stayed on or they would both be banished. His wife disliked me intensely, and for just a moment I thought she was going to protest, but she must have thought better of it and wisely kept her mouth shut. His former friends, regular customers who used to enjoy his company, had given him the nickname "mucho boozo" and begun disassociating themselves. He was offered a choice at work: rehab and a promotion or early retirement. He chose early retirement.

Now that I had a well-found boat, I could roam much further in the fall to pursue tuna. My crewman took a couple days off from UNH to make a trip because he really wanted to catch a tuna. I decided to head for the southwest corner of Middle Bank, roughly 60 miles from Seabrook. Middle Bank is shaped like a crescent moon stretching across the middle of Massachusetts Bay from roughly ten miles south of Gloucester to eight miles north of Provincetown. It is part of the same glacial deposit that made Cape Cod. The bank is covered with an astonishing array of life, all fed by sand eels, a small, eel-shaped fish that burrows in the sand. The southwest corner is the point at which the bank turns from running north to south to heading east to west. High quantities of bluefin feed here in September and October. Bill and I arrived in the middle of the night after a six-hour ride, and while a few boats had stayed out overnight on anchor, the section of the bank I wanted to fish was open. We anchored and went to sleep for the night.

We were awakened in the morning to vessels arriving from every direction, perhaps 150 in all. I explained how to fish and what was required when a fish hit. We put out the lines and then we waited, but not for long. Off in the distance, from the Cape Cod Bay side of the bank, you could see white water moving directly towards the fleet. These were tuna pushing bait. The tuna entered the fleet and pandemonium ensued. Boats were hooking fish left and right. Some had more than one fish on at once. Engines were being fired up, anchors let go, people screaming, just crazy. Then one of our lines went off. The fish ran under the boat, taking line out like I had never seen. We just barely got a second line tied on before the first basket was empty. Line was still smoking out. I was trying to work the basket down the rail and around the stern, but the line was moving so fast I could not grab it. The second basket was rapidly being emptied. I told Bill to cut the dart off the basket of line it was attached to and tie that basket to the second one. Of all the tuna I had caught, not one had ever entered a third basket of line, but this fish was going to do it. I also told Bill to tie a flotation ball to the anchor line and then drop it. This would hopefully get the boat to spin around so I could get the line to the stern before it got caught on the keel, propeller, or rudder. Halfway through basket number three, the fish slowed his attempted escape. Just dragging 1,800 feet of rope must have been tiring it out. We had been anchored in 120 feet of water, but 1,800 feet astern of us the water depth was about 70 feet.

I had never seen a fish run to the shallow water, so this was going to be a different type of battle. Surveying the angle of the line, and the fact that an anchor ball was tangled in it some thousand feet away, I decided to have Bill play the fish while I tried to back the boat through the fleet. This would hopefully allow us to retrieve much of the line, because if the fish made a big circle to head back to the deeper water, it would entangle multiple anchors, and most probably break the leader. Maneuvering the boat backwards is not easy, but gradually Bill gained back basket number three and most of number two. The entangled skiff anchor reached the boat, and I was able to easily untangle it from the tuna line. At this time the line went slack, and Bill thought the leader had broken. I told him to pull the line as fast as he could. I had seen this happen before. The fish turns and runs right back at you furiously shaking its head back and forth trying to get the hook to fall out. In this case, the maneuver was fatal. The slack line became entangled numerous times around the fish's tail making it difficult to swim. I took over the line from Bill and instructed him to rerig the harpoon line. I could feel the fish's tail beating furiously on the line and wanted the fish darted before it became untangled. The fish emerged from the depths tail first, which meant darting it was going to be a challenge. I turned the line back over to Bill, took a position about five feet up the rail and waited. When the tail emerged from the water, I threw the harpoon at an angle, and it struck reasonably close to the heart. The fish was exhausted and stunned, giving us a minute to place a strap over the tail and secure it firmly to a cleat. This was the first time I really had a chance to see the fish. This fish was monstrous, easily twelve feet long, with a huge girth, larger than any fish I had ever seen.

The boat was a tangled mess of rope, overturned baskets, and hooks. We towed the fish slowly back to our anchor and secured ourselves. Several hours later, we had the boat ship-shape, but we were both exhausted. I decided to haul anchor and tow the fish to Provincetown slowly. We arrived in Provincetown around four in the afternoon and found most of the commercial boats from our harbor and others tied up. The price of tuna had dropped, and they had decided not to fish, or more precisely had decided to spend the day in the local bars. Provincetown is a place where virtually anything goes, but the fleet was pushing the limits of the town's goodwill. We backed under a hoist, and were immediately met by the tuna buyer and the harbormaster, who noted Seabrook, NH, on the stern of the *Lady Marcia Ann*. He informed me in no uncertain terms that there would be no shenanigans tolerated. In return I informed him that all we wanted was to sell our fish, eat dinner, and go to sleep. The hoist was lowered, and an attempt was made to raise the fish from the water. The hoist was rated at one thousand pounds but it stalled out before the head was out of the water. With the assistance of three other men, we were able to put a gaff in the fish's mouth and bend the fish over the stern into the boat where it was lowered back down. The true size of the fish was now on full display and the pier became crowded with onlookers. The dealer threw down an extension cord and climbed aboard with an electric saw to cut off the head and remove the entrails. This would lighten the fish by about 20 percent, hopefully allowing it to be lifted onto the pier. Straps were placed around the fish as the tail was looking like it might rip off from the strain. The hoist groaned and barely moved but did finally lift the fish up onto the pier, where it was placed on a scale. The scale read 948 pounds, over 200 pounds more than my next-biggest fish. Adding back the estimated weight of what remained in the boat, this monster weighed around 1,140 pounds whole!

I was instructed to back into a slip by the harbormaster. We secured the boat, and I left Bill to start the grill for dinner while I went to the buyer's office to fill out paperwork. When I returned, Bill was grilling steaks and the sun was beginning to set. Suddenly there was a huge crash and the boat moved wildly knocking the grill onto the deck. I grabbed a bucket, dipped it in the harbor, and threw sea water on the hot coals that were about to set the deck on fire. Then I went forward to find out what the hell had happened. The boat had a pulpit made from an aluminum road sign with a plank bolted onto it for harpooning tuna. This stuck out about seventeen feet into the harbor and was about twelve feet above the water at its outer end. Apparently, a lobster boat from my harbor, manned by somebody under the influence of something, had tried to drive his boat under the pulpit. He chose poorly! The topside of the lobster boat was floating upside down in Provincetown harbor. Two exhaust pipes stuck up out of the bilge and everything from the dashboard up was gone. This was the final straw for the harbormaster. He ordered all the transient boats out of the harbor by sunrise the next day.

CHAPTER 12

A Fish Named Fred
and Family Ties

The wind had shifted into the northeast overnight, and as we departed Provincetown the radio was already alive with people complaining about the ride to the southwest corner. There is an area east of the tip of Cape Cod called "the race" which is well-marked on maritime charts with a warning to mariners about locally treacherous conditions when winds are either out of the northeast or southwest. Basically, as water tries to enter or exit Cape Cod Bay between the tip of Cape Cod and the southern end of Middle Bank, the tide runs up against the wind-driven waves creating a very steep, chaotic sea. One minute the ride was an unpleasant pound into a northeast sea, the next it was foaming mountains appearing in the spotlight and breaking over the bow in thunderous torrents.

I brought the boat to an idle and noticed we were still moving at over six knots as this invisible wall of tide pushed us to the northeast. Just before reaching the bank, the tide subsided just as quickly as it had started, and we resumed our normal speed. Many boats had turned around because of the weather, and only a few large draggers and some party boats were on the corner. We anchored and resumed our tuna-fishing routine. Once again around 9:00 a.m., fish were sighted off to the west, moving towards the bank. Their speed, however, was much quicker and it soon became apparent why. The tunas this morning were the bait, not the predator. They were being chased by at least two killer whales. This was an inspiring sight, as the whales with their distinct, three-foot-high dorsal fins slashed at the frenzied tuna. The tuna came into the fleet and two boats hooked fish, but the tuna kept right on going. One of the whales passed right down our side, and we had a bird's-eye view of their distinct white-and-black markings. The whale we saw was about twenty feet long and moving just as fast as the tuna. The tuna and the whales continued off to the east out of sight. In keeping with what old timers had said, there was not another tuna caught in that area for the rest of the fall. The whales had driven them away. After not marking a fish for five hours, I packed up and headed back to Seabrook. Bill needed to return to UNH, and I needed a new place to fish.

I had taken a job with a captain from Newburyport for the winter, as Rocky had hired a captain for the *Denise Ann* who had his own crew. I had never fished out of Newburyport, which did not have a large commercial fleet because the entrance to the harbor, the mouth of the Merrimack River, was dangerous and unpassable in an easterly wind. In fact, other than the Columbia River in Oregon, this was the most dangerous harbor entrance in the United States. The boat I was working on was a forty-foot party boat outfitted with a pot hauler and gillnet lifter for winter fishing. The boat was owned by Captain Fred, who was a local character steeped in Yankee traditions. If you wanted to cast someone in a movie to play the part of a Yankee fisherman, Fred was your man. Fred had wavy white hair, a long white beard, and wrinkled face that made him look like the old man of the sea. He always sported a wry smile and had endless witticisms, such as: "the big march on and the little sink," what he called the first rule of the road, and "don't get mad, get even," plus "10 percent for handling" when he was dealing with someone he did not like.

Fred's plan was to haul gillnets, sell the edible fish, and keep the unsaleable fish for lobster bait. He planned to haul the lobster traps every three days. Almost immediately, while still rigging the gear, my freshly retired father showed up demanding to go along. I called Fred aside, apologized, and explained the problem. It turned out Fred was a longtime friend of Bill W., also known as Alcoholics Anonymous, and saw my father as a very viable candidate. He let him ride along on the condition he attend a meeting when we returned. This he did every day all fall, and the process worked. My father was sober for the first time in my life. Fred let him come along, and every afternoon they attended a meeting. We were doing well on the cod, and the lobster traps produced a day's pay twice a week. I enjoyed the gillnetting, but hauling traps was not exciting. Basically, the trap came to the surface; you removed the lobsters and the occasional cod, winter flounder, or wolfish; put in fresh bait; and moved the trap to the back of the boat until the entire string of traps was hauled. Then, you reset the traps and moved on to the next string. The only drama was how many keeper lobsters you would get per trap. We gillnetted until early January, when the cod seemed to leave, and lobstered until late January when that slowed down as well.

Life had settled into a predictable pattern, seven months of party boat fishing, four months of gillnetting with February off.

In the spring of 1979, Fred and I resumed gillnetting. We had strings of nets on several humps east-northeast of the Merrimack River and one string on Skippers Rock about five miles south of the Isles of Shoals. Fishing was slow as the cod had not really shown up yet. There was one exciting event though. While hauling on Skippers, a very large white fish began to emerge as the hauler strained. It was an Atlantic sturgeon between ten-and-twelve-feet long, very feisty, and quite alive. Now sturgeon is one of the oldest fish from an evolutionary perspective. Their skin is made up of a series of bony plates; they do not have scales. Their mouth is small relative to their size and located on their ventral side. They root in the mud for worms and other invertebrates. Sturgeon swim into rivers to spawn and then go back to the ocean. They return to the same river to spawn year after year, an event called natal homing. They have well-developed eggs and are stripped for caviar in Europe. There is no current market for sturgeon in the US, so we untangled the fish and wrestled him over the side where it swam away. The next day, hauling the same net, what appeared to be the same fish arrived on deck again. This set off a lively discussion about whether it was, in fact, the same fish. I ended the discussion by grabbing a can of spray paint from down below and spray painting the captain's name, Fred, on the sturgeon's side. The paint did not stick very well, but there was a faint red name visible, when once again we released Fred alive over the side. The discussion was ended the next day when Fred appeared yet again and continued appearing daily for the next week. We finally moved the string of nets, as Fred was tough on the equipment. However, you would hear periodic chatter on the radio as fishermen reported catching a very large sturgeon that appeared to have Fred faintly visible on its side. Fred was probably the first celebrity fish!

Captain Fred had a lasting impact on my father, and he stayed sober for the rest of his life. This was a monumental achievement, which changed his relationship with the rest of the world. Probably because he was sober, his marriage to wife number three began coming to an end. Whether he just was not as much fun sober, or if sobriety made it painfully clear they had nothing in common, I cannot say, but by the end of the summer, he was single again. This made me the entertainment committee and began his tradition of fishing on whatever boat

I was running or working on for the rest of his life. I enjoyed having a normal father for the first time in my life, but having him with me constantly sometimes proved challenging.

Ellen and I had decided to start a family. She was concerned that the severe medical issues she had in high school and extending into her first year of college might have left her incapable of having children. One afternoon in the fall, I got home to find a note that she had gone to the local emergency room for unbearable pain. I drove to Exeter to find her admitted for an ectopic pregnancy. She was alternating between feeling disconsolate and white rage at what she considered a misdiagnosis. The next morning, she checked herself out against doctor's orders and drove back to Hampton. Ellen rifled through her college files until she found the phone number of the doctor who had treated her. After talking on the phone, she was summoned to Boston for an immediate examination by a gynecological colleague. The man took one look at her and categorically announced congratulations on being pregnant. He confirmed his diagnosis with an ultrasound and referred her to a local gynecologist in Newburyport for a full work up and prenatal counseling. Ellen had run the emotional gauntlet from deep despair to complete elation in one twenty-four-hour period. I always wondered how the people in Exeter had gotten the diagnosis completely wrong, but medicine is part art and part science, which is why second opinions are so important. Barring unforeseen consequences, Ellen would give birth in early April.

Within weeks Ellen had resigned her biologist position, as she had been working with an assortment of chemicals that might cause problems for a developing fetus. She did, however, continue to teach figure skating almost up to the time of giving birth. We attended birthing classes and practiced breathing on the living room floor. Ellen radiated joy, even through the first trimester, when holding down food was a daily challenge. I continued fishing with Fred but had begun to look at buying my own gillnet boat. Fred had announced he was selling the *Brenda* after the fall season to trade up to a bigger party boat. His new boat would not be rigged for commercial fishing, so I would need to find another job. Two fishermen from Newburyport had twenty-four-foot fiberglass skiffs with gillnet lifters mounted on the bow. They fished by themselves and did quite well. I decided to investigate having one of these boats built for myself.

Between Thanksgiving and Christmas, I decided to meet the boat builder, who lived in Eastport, Maine. The trip would take about nine hours each way and require staying overnight. I had never been east of Rockland and had neglected to check into the availability of motels. Poor planning on my part. Ellen and I left at daylight and arrived after dark in Eastport. We both noticed there were no open motels or restaurants along Route 1 for the last one hundred miles of the trip. The builder, whose name was George, and his wife both knew this and had made plans to feed us a salt pollack boiled dinner with their family and set up one of the units in their small motel for us to stay in overnight. While George and I discussed the specifics of building the boat, Ellen learned how salt pollack was prepared. Apparently, you soak the fish for about twelve hours to pull the salt out of the fish, but because they did not know with certainty that we were coming until we called from a gas station, the fish had only been soaked for four hours. Dinner was the fish, potatoes, cabbage, carrots, and a couple other root vegetables. As promised, the fish was salty but tasty, how salty would not become clear until the middle of the night. After dinner we were each given two bottles of water to brush our teeth and drink as needed. The water had been turned on in the unit for the toilet, but the family recommended not drinking the water. We were provided with blankets and quilts as the unit had no heat. By one in the morning, we had exhausted our water supply and both felt like we were in some Arctic version of the Sahara Desert. We were both shaking almost uncontrollably, whether from the salt, the cold, or both I could not say. Ellen and I seriously considered getting in the car and leaving but thought that would be rude given the hospitality we had been extended. Luckily, the family were early risers, and we were admitted to the house to get warm well before daylight. We had a quick breakfast and hit the road for the trip back to Hampton. We drove about seventy miles before finding an open convenience store where we purchased a six pack of water. Neither one of us had ever been that thirsty in our lives. The finished boat would be trailered to Portland, Maine, at the end of February to have the hydraulics and lifter installed by a machine shop.

George called near the end of February to inform me he would be delivering my boat to Portland around nine the next morning. I arrived at the appointed hour to find George and his family in a pickup truck with the boat in tow. The shop was located on a steep, cobble-stoned street which ran perpendicular to the Portland Harbor waterfront. George exited the truck followed by an empty rum bottle which rolled down the street. Apparently, the trip had been fueled by rum and black coffee, both of which had been fully consumed. George looked a little the worse for wear and was eager to unhook the trailer and head to their hotel for a nap. I looked at the icy street and had visions of the news headline to follow: "Boat on Trailer Takes Wild Ride down Portland Street! Details at Eleven." I suggested we wait for the machine-shop owner to find out if this was where the boat should be left. He arrived shortly thereafter and produced some chocks to secure the trailer wheels. George and his family were staying in Portland for their yearly vacation and would trailer the boat to Seabrook when the hydraulics were finished. Five days later, the completed boat was delivered and launched at the boat ramp in Seabrook. We spent the afternoon riding up and down the harbor making sure everything worked as planned. Both were duly satisfied, final payment was made, and George headed back to Eastport. A late-winter cold front swept through, and the temperature dropped close to zero. My new boat, named *Ellen Diane* after my wife, spent its first night in the water frozen in late-winter ice.

New boat, new baby! The first F/V *Ellen Diane*. David standing on the deck holding his eleven-month-old son, Eric. Note the gillnet hauler on the bow. The Eastman's dock, Seabrook Harbor, March 1981.

The weather soon broke, and gillnets were loaded on board and set out near the Isles of Shoals in some broken bottom. Cod were there feeding on shrimp, and the catch was perfect for my little boat, about five to six hundred pounds per day. On good weather days, I would haul the nets; on windy ones, I would get the *Lady Marcia Ann* ready to go back in the water. April 5 was windy but warm, and I put in a long day painting in the boatyard. I got home about six, ate dinner, and prepared to go to bed around eight. Ellen was walking up the stairs ahead of me when we both heard a loud pop, like someone breaking a small balloon. Ellen turned calmly and said I should start the car and get her suitcase because her water had just broken. Thirty minutes later, we were at the hospital in Newburyport, where Ellen was admitted and made comfortable. For the next four hours nothing much happened, and I dozed off between contractions in a chair next to Ellen's bed. After midnight, contractions became much stronger and intense. I did my best to follow the training we had been given. Around 5:00 a.m., the nurses told me to go to the cafeteria for breakfast and some coffee. I was assured no baby would be born in my absence. Duly fortified with caffeine and a plate of scrambled eggs, I returned to find the birthing process had kicked into high gear. Even though we had been fully educated on what to expect, the process was much more intense than I had expected. The pain Ellen was enduring made me sick to my stomach, and the scrambled eggs almost made an unwanted appearance. A little after 8:00 a.m. on Easter Sunday 1980, Eric David Goethel made his appearance in the world. I cried, both at the miracle of birth and because I was relieved Ellen's pain would now subside. The doctor and nurses placed Eric on Ellen's chest for the first time. The look of complete rapture is something I will never forget and still brings tears to my eyes all these years later. We were now a family.

CHAPTER 13

One Fish, Two Fish

From here on in, sleep deprivation was the name of the game. Newborns do not sleep much anyway, but Eric had raised fussing to an art form. Initially, he lost weight and always seemed hungry. This was not unexpected, but after several weeks the pediatrician decided to augment his regimen with formula. After trying several brands, one that was soy-based seemed to do the trick. The fussing subsided and Eric began to sleep a few hours at a time. Ellen and I were both totally bereft of sleep. I worked fishing every day and began to rely much more on the crew. Luckily, I had good crewmen who could handle the extra duty. As a business, Eastman's had decided to no longer let customers bring alcohol on the boats. Each boat sold beer and tonic. The hope was that the added cost would deter excess consumption, and the revenue helped the crew make more money. My crewman, Bill, had a Volkswagen Beetle. In a 1980s version of seeing how many college kids you could fit in a telephone booth, he had found he could fit twenty-two cases of beer in his car and still drive it down the road. While most of the beer was delivered by a distributor, in an especially thirsty week where we might run out, Bill would hit the local discount supermarket and load up. I am sure the checkout people wondered what kind of party this young man was throwing with multiple carts full of beer.

Ellen would bring Eric down when I was in at noontime, and we would eat lunch together. I would try to read to him after dinner to get him sleepy, but often Ellen would find I had read myself to sleep, and Eric was still wide awake. I still have Dr. Seuss's *One Fish, Two Fish, Red Fish, Blue Fish* committed to memory. Ellen's parents came to visit for a few weeks, giving Ellen some much needed rest and assistance. The year passed very fast. By winter, Eric was spending a great deal of time in a walker. For those that do not know this device, it was round with four wheels and a tray. The baby sat on a seat in the center and his legs touched the floor. He could move the device around by moving his feet, thus preparing him for walking. Eric never crawled; he went from sitting up to walking. He was incredibly inquisitive, and every part of the downstairs had to be completely childproofed. Childproofing, however, proved to be a relative term. What hardware stores and childcare books said constituted child-proof were just speedbumps in Eric's ability to outwit them. In early 1981, Eric somehow opened the cellar door, the handle of which we did not think he could reach let alone turn, opened the child gate behind it, and hurled himself and his walker down the cellar stairs. The device kept him from serious injury, but he cracked his wrist and needed a cast. He came home from the hospital with a cast from the base of his elbow to the top of his hand. He was quite proud of the cast and showed it to everyone he met. I installed a deadbolt two feet above the door handle, but he was still not deterred. About six months later, he did it again. This time the hospital was not amused. Ellen and I were put in separate rooms and grilled, as I am sure the staff suspected child abuse. Eric explained to a nurse how he had outwitted all the locks and gates. Eric had picked up the nickname at the pier of Tasmanian Devil for his ability to get into everything conceivable and for the fact he never stood still for a second, but he called him-self by a different moniker, Happy Eric!

In the fall of 1981, I added lobster traps to my little gillnetter's to-do list. Fred wanted to sell the traps he had to get them out of his yard, and I ordered fifty more to give me enough to make lobstering worthwhile. Lobster fishermen have territories and starting in the fishery always involves some risk. Even though I fished out of Seabrook, I lived in Hampton, so I set where the Hampton boats fished. The area I chose was considered open territory, that is, people from Seabrook and Hampton both fished there. If you did not set over someone, you should be accepted. The first month and a half, I had no issues. Then I started losing traps. I fished five traps tied in a string called a trawl. I would lose one or two traps off an end of a string. Another boat had showed up, run by a person everyone in the harbor considered trouble. The rumor mill went to work, and they had their culprit. Lobstering was about done, so I took my remaining traps and stacked them in the yard for the winter. I figured the problem would be gone by spring. In early March, I resumed gillnetting. One cold afternoon, I walked up the pier in boots and rain pants wearing two sweaters and a wool shirt. I looked like the Pillsbury dough boy and could move just about as well. There I was confronted by this character, who led with an unexpected fist. I ducked and only received a glancing blow but lost my balance and fell on the pier. I soon found out why you cannot fight in a strait jacket. I could not move with all this stuff on. I covered up and took a beating, being kicked in the ribs several times and in a final blow, stomped in the face. There were plenty of onlookers, but no interveners. Someone, however, had called the cops. The sound of sirens put an end to this episode, but as he ran off, he threatened to burn my house and kill my wife and son. I gave a statement to the police who summoned me as a witness as they went off to arrest the offender. I called Ellen and had her take me for x-rays of my ribs and face. For two weeks afterward, I had black and blue indentations from the sole of that shoe on my face.

The justice system proved useless, as assault carried a thirty-day suspended sentence. My wife was friends with the wife of the Hampton police chief. He stopped by and told me in no uncertain terms to buy a gun, but Ellen wanted no part of a gun in the house with a two-year-old who could defeat every child safety device known to man. Our chief despised drug dealers and gave his officers orders for a stop and frisk the man if he was spotted in Hampton. The problem soon resolved itself as the individual was arrested for distribution in a neighboring town and sentenced to five years. On release, he promptly overdosed on cocaine, making the solution permanent, but I did buy a Louisville slugger that still resides under my bed to deal with intruders. An interesting side note is that several years later, Fish and Game contacted me about several traps with my tags washing up after a storm. They wanted to know if I had sold any gear to another person from Seabrook whose tags were also in the traps. That person had recently died, so no charges could be leveled, but I always wondered if he had stolen them from the drug dealer or stole them from me directly. Either way, I had moved on; lobstering was not worth the hassle for the money involved.

The year 1982 was not off to an auspicious start. Matters were made worse when my mentor, Bill, now suffering from terrible Alzheimer's, walked off the end of the pier on his return from Florida. Bill was now almost completely nonverbal, and no one knew whether he just forgot where the edge of the float was or if this was an attempted suicide. Either way, a stint in the March harbor water nearly killed him, and he had to be placed in a nursing home. He would die soon after. Bill and Lil were the glue that held Eastman's together. Even though

Bill had not been active for four years, his presence sitting in a rocking chair in the ticket office seem to have a calming effect. Lil was like a mother superior, counseling all the women, doing the books, cajoling the regular customers, and keeping a weather eye on the kids and crew. Things just seemed different, or maybe this particular year was a harbinger of change.

Gillnetting in April and May was extraordinarily good. I had to bring in some of the nets because I was catching more fish than my little boat could carry. One day in May, I called the half-day boat and asked them to come by. The boat only had a few people on board because the kids were still in school, and this was a weekday trip. I filled the deck on the portside almost to the rail with cod, told the customers to each take a fish for their lost fishing time and returned to hauling. I filled the *Ellen Diane* again with cod and went in to clean both boats full of fish. When I was through unloading both boats, I had tallied about forty-five hundred pounds of cod and a few haddock.

Word gets around quickly in a small community when someone has a large catch. The next morning, at the coffee shop, across from Eastman's ticket office, I was chatting with the owner, Don, who also had a gillnet boat, and Marvin, known to everyone as Red because of his shock of carrot-red hair, about the previous day's catch. Marvin was just naturally friendly, always going out of his way to help people, and was someone you just could not help but like. He was in the process of buying out the elder statesman of the harbor, Frank, who owned a dragger. Frank had been a pilot in World War II, smoked a pipe, always wore boots, even on land, and just looked like a fisherman. He was quiet and seldom spoke, but when he did speak, you should be listening. I had made a few trips with Frank just for fun to see how dragging was done. Basically, you had a net wound on a hydraulic net reel. You closed the end of the net with a bronze clip called a tripper, let the net out and the cables that were attached to the net—called ground gear—allowed the net to open. These were then attached to a wooden trawl door on each side of the boat. The function of the doors was to spread the mouth of the net open when the trawl wire was paid out from a winch containing two drums of steel cable. The net was towed over the bottom for a period, after which it would be hauled in and the captured fish dumped on deck. Frank liked me and had offered a few pointers over the years since I bought the gillnetter. I knew he worried about me and thought I took chances carrying too many fish on my little boat. Frank fished for flounders, probably because he fished alone and flounders did not need to be cleaned. Frank was concerned I had brought all those fish in on the *Ellen Diane*, which would have been impossible. I told him how I loaded up the party boat. He smiled and I could see he admired the ingenuity. We chatted a bit more before Marvin and Frank left for their fishing trip.

By the middle of June, the cod had thinned out and the spiny dogfish had arrived. Dogfish are a small schooling shark with two sharp spines, one behind each dorsal fin. At this time, they had no market value and clogged the nets in enormous numbers. They were the chief reason you could not make a living year-round with gillnets. This spring they were particularly savage. I would work all day removing ten–fifteen thousand pounds of dogs from one string of nets so I could bring the nets in. If I was lucky, I would get enough fish to buy the gas. Dogfish have scales like sandpaper and after a day of hauling you would have bites, spine wounds, and no skin left on your fingers. This would be the end of gillnetting until the dogs left in October.

The summer of 1982 produced the strangest collection of characters who fished on the party boats in my career. The first incident was a charter from a bar on Salisbury Beach. This was an afternoon half-day charter, as this crew were not early risers. The group were mixed male and female, probably bartenders and bouncers on the male side and cocktail waitresses and perhaps a few strippers from some of the other clubs in Salisbury. At any rate, they fished intently and were generally well-behaved until near the end of the trip. Suddenly one of the ladies removed her clothes, stood on a bench on the stern, and began performing various sex acts with some of the men. Now I had seen people do a lot of strange things on these boats in my career, but this was a definite first. They

Capt. Dave and a large cod after a successful trip aboard the *Lady Marcia Ann* for Eastman's Fishing Parties. Late 1970s.

were not, however, doing anything that interfered with the safe operation of the boat. I put the story on the radio, probably not the best idea in retrospect, but since they obviously had no shame, why not entertain the fleet? The nearest boat was a large, all-day boat run by my friend Rocky. The boat had the pilothouse towards the rear of the of the boat with a large, open bow. I guess his plan had been to pass down one side of my boat, as his boat was faster. However, when all sixty odd men on board rushed to the bow to get a better view, the bow dropped in the water and the rudder blades which provide steering came partially out of the water, and Rocky lost control of the boat in my wake. The boat slashed wildly across my stern, and the bowsprit—a heavy beam that protrudes from the bow and holds the anchor—passed over my stern, nearly cleaning the young lady off the bench. Rocky backed one engine down hard to regain control and ran off about a half mile to make sure no other issues occurred. I guess that was the impetus for the end of the show, because when I sent the crewman back to tell them they needed to be clothed on arrival at the pier, they were all just sitting around drinking beer like nothing had happened.

Drug use was rampant, and we had almost daily issues with pot smokers. The biggest problem was they always seemed to try and get on the roof of the cabin, I guess to hide their illegal activity. The Coast Guard has very strict rules about keeping people inside the rails of the vessel and a zero tolerance for illegal drugs. This meant the crewman had to haul them off the roof and feed their illegal stash to the seagulls. This was always met with some version of: "Bummer man, I was just trying to have a little fun and catch a buzz." Dope smokers were annoying, but generally docile. However, this summer every kind of dangerous drug from meth to coke to PCP seemed to be circulating in the male population. Meth made them crazy; coke made them intensely paranoid, and PCP made them paranoid and insanely strong. I had an incident with a man who I assumed took coke and started having paranoid flashbacks to Vietnam. I had a large group of people with Asian ancestry on board, and as this guy studied them more and more, it became obvious some kind of problem would occur. I assigned the crewman to watch him, but without warning, just after passing under the Hampton Bridge on a strong outgoing tide, he jumped overboard. The crewman immediately told me, and after passing far enough upriver to complete a U-turn safely, we went back to retrieve him. I explained to Gary how I would run past him down river, turn around, and approach the swimmer up-tide. I wanted Gary stationed just behind the wheelhouse door to grab the guy if he extended his arm. This was dangerous; he could pass under my boat and get ground up in the propeller, be run over by another boater, or just plain drown. The man did nearly drown when the eddy under the bridge sucked him under, but he resurfaced. We completed the maneuver, and as we approached the guy started waving one arm. Apparently, the trip under the bridge had convinced him he would not win an Olympic swimming contest. Gary was a big, strong college man and pulled the swimmer with one arm over the rail. As he did, he pushed the guy's arm down sharply over the rail's edge. There was a sound like someone stepping on a crate of eggs as the guy flopped into the boat screaming. The guy looked up at Gary who calmly stated that "You won't be going swimming anymore with one flipper, will you?"

I had a morning off in early August. Ellen brought Eric down to spend some time with me. Eric delighted in turning the big captain's wheel in the wheelhouse and playing captain. He could not hurt the steering, and this kept him happily playing while I worked on the boat. Eric was now two and a half and was always observing and learning how things operated. Unbeknownst to me, he had figured out the throttle, the shift, and how to start the boat. The starter button was on the port-side instrument panel and consisted of a spring-loaded switch that had to be pushed in to complete the circuit to start the boat. The throttle and shift were on the starboard side. Eric had put the shift in gear and jammed the throttle wide open. I was working in the cabin when I heard the engine scream to life. This engine was equipped with an Allison transmission, which took about five seconds to pump up hydraulic pressure to engage the clutch. Fishermen used to call it

Customers showing off their catch in front of East-man's ticket office, 1982. Capt. David Goethel, *second from the right.*

the Allison lag. I reached the wheelhouse and pulled the throttle back to idle just as the clutch engaged. The boat shot forward but did not rip the pier apart as the engine returned to idle. Eric was beaming, telling me he was driving the boat just like dad! We had a long talk about needing to be an adult to start a boat, but from then on, I shut off the battery power with the main switches just to make sure Eric never took an unauthorized ride.

The end to my party-boat career came on a late August afternoon. Brad had four men on his boat who had used PCP and decided to keep a bunch of life jackets as souvenirs. PCP is a horse tranquilizer, but when ingested by humans makes them intensely paranoid and incapable of feeling pain. Basically, you could hit one of these guys with a baseball bat, and if you did not kill him or knock him out, he would just act like nothing happened. Brad radioed the other boats, both so they could land first to get their people out of the parking area and to have overwhelming back up to take back the purloined life jackets. The police were also called, and a cruiser was in the parking lot. Brad was a big guy, having played goalie in college hockey, and no one really expected a big problem. Brad told the apparent leader, who had a life jacket on each arm, to return the equipment. The man dropped the jackets and punched Brad twice in the face breaking his nose and jaw. Then all hell broke loose. Several of us were trying to ride these guys into the ground, but it was hopeless. The cops never got out of the cruiser. We got flung around the way broncos toss rodeo riders. The offenders gained their vehicle and sped off towards the Massachusetts border with the cruiser in pursuit. The pursuit was broken off when the thieves sped into Salisbury. Meanwhile, the parking lot looked like an apocalyptic scene. Brad was knocked out, and an ambulance was on the way. The rest of us had cuts, bruises, bloody noses, and black eyes, but nothing serious. When I walked through the door of my house an hour later, Ellen just said, "What now?" We both agreed this was not how you raise a family. We had a son and hoped to try for another child soon. No one else had dads who came home from work looking like a refugee from worldwide wrestling. After seeing Brad in the hospital the next day, unable to speak with his jaw wired shut, I made up my mind my party-boat career was over.

CHAPTER 14

Building a Boat

I did not make my exit plans known right away. Perhaps, in retrospect, that was unfair to the other owners. I wanted the season to end, haul the boats, and store them for the winter; basically, to complete the year's obligations. My plan was to gillnet through December and then put the *Ellen Diane* on the market. I told the other owners of my desire to exit the business at the end of October. They could buy me out or I could sell on the open market. The owners decided to sell the boat I was running to an outside investor and use the proceeds to buy me out. With that money plus the money from the sale of my gillnetter, I would have a substantial down payment for a dragger. Red had purchased Frank's dragger and renamed the boat for his daughter, Alexis Marie. I made several trips with Red that fall, both to get a better handle on setting and retrieving the gear and to discuss what he would look for in buying a dragger of similar size. I knew a lot about biology and boat maintenance, but Red knew much more about hydraulics, electrical systems, mechanical systems, and plumbing. We were friends and his advice was very useful. Draggers, from a system point of view, are much more complicated than either gillnetters or party boats. I contacted a boat broker and received a list of used boats in my size and price range. I spent the windy days in November driving the length of the New England coast looking at boats. Rapidly, it became apparent that the cost of a used boat combined with an estimated cost of converting it to what would work for me was about equal to the price of a new boat. With a new boat, I could build it exactly as I wanted. Of the used boats I looked at, a model called a Stanley 44 seemed the best fit for my ideas. My next stop was a government loan expert. Interest rates were through the roof, and a commercial boat loan had an interest rate of 18–19 percent. However, the government would guarantee your loan if you bought what they wanted and that would drop the interest rate to about 8 percent. They also had an investment tax credit which reduced your actual taxes paid by 10 percent of the value of the boat. So, for example, if you spent $200,000 for a boat, your taxes would be reduced by $20,000.

I soon learned that the government only wanted large boats. The loan expert tried very earnestly to talk me into building a ninety-foot steel dragger, but I wanted a small boat. He said he could not help and sent me back to get a private loan. The federal government's selective funding for oversized, expensive fishing boats was perhaps their most disastrous policy that undermined the long-term sustainability of the fisheries. The government had clearly learned nothing from what the large foreign vessels had done to fish stocks. The attitude was we are a first world country, we need a first world fleet, which to bureaucrats meant "bigger is better." The problem was compounded by the tax credit, which soon had every high-income earner building a boat, even if they could not drive it out of the boatyard on launch day. This gross overcapitalization of the New England fleet would ruin the fisheries and carries on over forty years later. Government, to this day, still refuses to acknowledge its role in the overfishing of New England groundfish.

I had little trouble in finding a banker and a bank. The man was impressed with both my knowledge of the fisheries and my business plan, which laid out poor, average, and good seasons. I think his board of directors was impressed with my proposed 30 percent down payment when most people were putting up 10 percent. Either way, I shaved 3 percent off the interest rate and had a total budget of $165,000.

Armed with my line of credit, I headed for the John Williams boat company in Hall Quarry, Maine, on Mount Desert Island, maker of the Stanley 44. The high interest rates had put a crimp in boatbuilding, and the yard did not have a winter project. We spent two days drawing up detailed plans. The yard would build the boat, provide the engine, wiring, and mounts for the winches and A frame. An electronics shop in Salisbury, Massachusetts, would install and wire the electronics, and after launch, a shipbuilding company in Portsmouth, New Hampshire, would provide the hydraulics, A frame, and winches. The total was $160,000; I had $5,000 left for nets and trawl doors. Everything was on a cost-plus basis and inflation was the highest since World War II. This would be close.

I had a busy winter ahead. For starters, I needed to learn how to mend dragger twine. If you are going to tow a net, sooner or later you are going to rip the net and you must know how to put it back together. I signed up for a twine course at Massachusetts Maritime Academy in early February. The course was two days and after several hours it became apparent to both the instructor and me that I was a slow learner. Twine is made for right-handed people and so are the knots to fix it. I am left-handed, and when the instructor tried to sew left-handed, he was no more successful than I was. Over lunch, he invented left-handed knots. The motions are awkward and basically mirror images of right-handed knots, but they work for me. My mending is ugly, but it holds; however, I still cannot teach right-handed people to sew.

Most of the construction planning was done by telephone, but I tried to visit Mount Desert Island every two to three weeks. The ride was over six hours each way, so with a couple hours of discussions these were long days. My son, Eric, kept begging to go along, so one day I got him up at 4:00 a.m., fed him while the truck warmed up, and finally strapped him into his car seat. I figured we would drive about an hour before the "are we there yet" would start, but he seemed genuinely happy. We stopped for bathroom breaks a few times, ate lunch at a McDonald's in Ellsworth, and arrived a little after noon in Hall Quarry. Eric was duly impressed with the hull, which did look large in the boatshed, but he soon resorted to his Tasmanian Devil form when we got up into the boat to measure and discuss placement of numerous items. The deck was not installed yet, and Eric could get seriously hurt. His interest was piqued by the rudder compartment, called a lazarette, because the rudder post had a bronze arm bolted on which he could push back and forth. Also, there was lots of sawdust to crawl in and throw about. He stayed happily occupied while we measured and talked, and the builder took copious notes. I retrieved him when we were finished and realized he was a total mess. Many of the yard workers had kids, and we had a collective effort to clean him up for the ride home. Eric was grinning ear to ear; this had been the best day ever.

In mid-March, I interviewed a net builder in Gloucester, or perhaps I should say he interviewed me. Before he would build nets for me, he had three conditions: first, I would drive him around to the various marine supply stores and pay for everything he bought; second, I would produce two bottles of Canadian whiskey at the completion of the nets; and third, I would never, ever, use his nets in Ipswich Bay at night. I agreed to his conditions, although I had no idea what the significance of the last two demands were. I judged Dan to be in his eighties and clearly deeply set in his ways, but his nets had a great reputation for catching fish and I was glad he would take me on as a customer. Dan had been a twine man on offshore boats, having fished the offshore waters of the northwest Atlantic from the Grand Banks to Cape Henry, Virginia. The job of the twine man was to keep the nets fishing; there were no marine supply stores hundreds of miles at sea. Dan was self-taught and designed his own nets. Despite his age, he was tough as nails and could work circles around men a quarter his age. Our first job was to buy the supplies to build the nets at the numerous gear stores in Gloucester. I was paying for everything, but Dan acted like these suppliers were draining his life savings. I drove him from store to store where he bartered with the suppliers. Dan had immigrated to the United States as a child and grown-up poor. He knew the real value of every item we bought and had no intention of paying one cent more than necessary. The discussions were animated, some might say heated, but he got what he wanted at a price he thought fair. He saved me thousands of dollars which, by this time, I really needed. I was sent home with instructions to come back at nine to begin net building in his basement and backyard. When I arrived the next day, I found him in the backyard with one net nearly finished. When I asked how he got so much done, he replied he could not sleep and had started cutting out the net pieces at 2:00 a.m. in his basement. I was told to get to work filling needles, which he used faster than I could fill them. This pattern repeated itself each day. He did most of the twine cut out and construction before I arrived. We would then spend the day driving around buying supplies and the afternoons building the sweeps that attached to the bottom of the nets. Dan built two complete nets with sweeps in less than a week. He wanted cash for his labor. I told him I needed a number to get the money. The number he requested was ridiculously low. I tried to bargain him up, but that was hopeless. His response was that he was old and did this to help people, not for the money. I was instructed to show up tomorrow with two bottles of Canadian whiskey.

I arrived the next day with the two requested bottles in tow. We went out into his backyard where the two nets sat in piles about four-feet cubed. Dan opened the first bottle and began to chant in Italian or maybe Sicilian or Latin, I was not sure. He carefully moved around each net sprinkling whiskey over every square inch of each net until the bottle was empty. He told me that now the nets were properly blessed and would catch fish. We then proceeded to drag each net out to the street and load them by hand in my truck. Keep in mind, one net had a roller sweep on it, which weighed over a thousand pounds. I was worried he would have a heart attack, but by the time we were done he was barely winded, and I was more worried about myself. I was then invited into the kitchen, where, under the watchful gaze of a picture of President John F. Kennedy over the entrance and an icon of the Virgin Mary, we drank to my future success. Dan had one further order. I was to never use these nets to fish Ipswich Bay at night. This was not a request. The Gloucester fishermen had learned years ago that fishing at night meant no cod five years later. This edict was enforced in a crude, but efficient way. The boats were all family-operated vessels and being shunned by the family was the least of your problems. The first warning was a knife in a buoy. The second was your electronics piled on the fish-hold cover. The final warning, should you be stupid enough, was your boat on the bottom. The fishermen had deduced a secret that science would not learn until the mid-2000s. Cod and haddock both spawn under zero light conditions, such as a new moon or on cloudy or foggy nights. Spawning is a ritualistic fishing behavior which can be disturbed by the noise of fishing and the glare of bright deck lights. Fishermen confirmed this fact repeatedly to federal officials, but, as always, they were ignored. Cod in Ipswich Bay never recovered when managers closed Georges Bank in the mid-1990s to rebuild the Georges Bank cod stock and forced the trip-boat fleet into inshore waters, where they fished day and night. These boats, derisively referred to locally as George Bush's thousand points of light, for their bright nighttime deck lights, wiped out the cod in two years, but in fairness, those large, offshore boats had been provided no fishing alternative. Driving home through Hampton Beach in the early evening, looking out on the lights twinkling eight miles offshore wiping out the spawning cod, I could not help but think how government officials ignored men like Dan who had six decades of fishing experience and managed to consistently snatch defeat from the jaws of victory in managing the fisheries.

By mid-April, my new *Ellen Diane* was nearing completion. The boatyard crew was led by Lyford Stanley, the man who had designed the mold for the hull. Lyford was an old-time Maine, wooden-boat builder who had reluctantly read the tea leaves of change and converted to working with fiberglass. My boat was a hybrid: solid-glass hull, but deck, focsle, wheelhouse, and rails made of wood which was then fiber-glassed over. I remember watching him cut out a sawn frame one day. He was working with a piece of oak a full two inches thick, fourteen inches wide and about fourteen feet long. The frame was for the focsle foredeck and needed a compound curve so the deck would properly shed water. The wood was clamped to two sawhorses, and Lyford made pencil marks on each end for where the cut should start and end. At this point, most carpenters would scribe the curve with a pencil and measuring tools, but not Lyford. He started his circular saw and started cutting and walking. The saw whined and complained, as oak is a tough hardwood, but he just kept walking until the cut was complete. The curve was perfect; you just do not find those skills anymore. The entire crew was dedicated to their craft and added a number of yacht-like touches to the wheelhouse and focsle. This took time, and time was money I was rapidly running out of, as inflation had eaten into costs. Realizing this, the entire crew came in and worked two weekends on their own time. The *Ellen Diane* was a representation of their talents to the outside world. They had a desire to

Eric, age three, supervising the construction of the dragger *Ellen Diane*, while being watched by Grandpa Jack. Jock Williams' boatyard, Hall Quarry, Mount Desert, ME, 1983.

show off their craftmanship to others. I still find it hard to put into words; the men were like artists who expressed themselves through their finished product.

Launch was set for a Friday in late April. Hall Quarry is located on Soames Sound, one of the few natural fjords in the United States outside of Alaska. Sandwiches and drinks were provided, as a new boat launch attracted the attention of the entire town. Ellen and Eric rode up with me, and Captain Fred came from Newburyport. He had graciously offered to ride down the Maine coast with me to Portsmouth. Ellen did the customary honors of smashing the bottle of champagne on the bow, the travel lift was fired up, and the *Ellen Diane* got her first taste of salt water. The yard crew swarmed aboard to check all the hundreds of things that might go wrong, but after about an hour pronounced everything shipshape and fired up the engine. Another period elapsed while the engine and transmission were duly checked out. Finally, the time had come to go for a ride. The first ride would just be the yard crew and me, as there were still numerous items to check and to measure in all compartments. We headed off down Soames Sound, slowly advancing the throttle and taking measurements. Once we reached the engine-maker's recommended speed for continuous duty operation, we used the LORAN to calculate speed over the water. We were going along at thirteen and a half knots, a very healthy speed for a semi-displacement dragger. I also knew the speed would drop once the dragging gear was installed and the fuel tanks were filled. Everyone was pleased, and I spent the afternoon giving people rides up and down the sound. The yard owner requested the boat stay for one more day so they could clean up the equipment still on board and recheck the propeller shaft alignment after the boat had sat in the water overnight.

I also had to officially take ownership of the *Ellen Diane*, which meant settling the bill. The yard owner produced a fifty-one-page itemized bill, which, when totaled, ran slightly less than fifteen thousand dollars over budget. The yard owner and I were now in an awkward position. I had exhausted my credit line at the bank and had nothing else to give. The cost overrun was no one's fault. The chief items responsible were external to anyone's control. Maine had changed its workmen's compensation laws mid construction, and the bank had required the yard take out a builder's risk-insurance policy. These two items accounted for most of the extra cost. We worked out an agreement that I would forward all extra money earned from fishing until the debt was paid. I honored this gentlemen's agreement, and by mid-September the yard was fully paid.

The *Ellen Diane* said goodbye to Soames Sound early on Sunday morning. Fred and I had one extra passenger, as one of the yard crew, a man named Charlie, had volunteered to ride down the coast. Charlie was a welcome addition as he had a good working knowledge of all the boat's systems. I am sure readers wonder what could go wrong with all new equipment in a brand-new boat, but the fact is, testing new equipment on a bench or in a shop is much different than pounding into a big sea. Eastern Maine is truly beautiful with its many islands and peninsulas running down to the sea. Most of the first day was spent in sheltered waters and uneventful. Our first taste of the open ocean was crossing the mouth of Penobscot Bay. The wind was fresh out of the south and we were steaming more or less southwest. The boat handled the sea well throwing sheets of water over the roof and occasionally burying the port bow in a big sea. There were no leaks in the focsle and only a minor leak in the corner of one window in the wheelhouse. Just short of the southern side of Penobscot Bay, the engine speed began to fluctuate. We all recognized this as a fuel issue and decided to put into the island harbor of Spruce Head. Charlie went to work on taking the fuel system apart and had soon located the issue. Detroit diesels have a high rate of fuel pumped through the engine. Some of the fuel is burned and the rest is used to cool the engine heads and passed back to the tanks by return lines. The tanks were made of fiberglass and had been cleaned after construction by pumping fuel through a filter from tank to tank. There is a big difference, however, between pumping fuel under controlled conditions and pumping the same fuel out of tanks being hurled up and down and side to side. The fuel filter was full of fiberglass hairs, and a fitting that stepped the fuel line down from three-quarters to five-eighths inch as the fuel entered the filter had a plug of fiberglass hair like a hairball in your shower drain and a small chip of fiberglass resin in its center. Cleared of these obstacles, the engine ran normally. However, we needed more filters before setting out again, and none were available on a Sunday. A local man gave us a ride to the only store, where we purchased what passed for food, i.e., bags of junk food, that we could consume on the boat. We could purchase fuel filters in the morning.

The next day, the wind was smoking from the southwest. Another man stopped by to say hello and offer us a ride to the store. People really went out of their way to look out for these three marooned strangers. Tuesday dawned with the same howling wind. This was getting old fast. I asked Charlie if he wanted to call home and have someone come pick him up as he was missing work, but he declined. Charlie seemed to relish the adventure. Finally on Wednesday, armed with a case of filters, after a set of heartfelt thanks to all the local people who had helped us out, we set out for Portsmouth. The wind was still fresh out of the southwest and the ride was not pleasant, but we only had to stop twice to remove fiberglass. We arrived in Portsmouth near sunset, tired and fragrant, and, at least for me, glad to be back in New Hampshire. Charlie stayed at my house overnight and the following morning Fred, Charlie, and I drove up to Hall Quarry so Fred could retrieve his truck and Charlie could return to his family.

CHAPTER 15

Good Business Relationships

The *Ellen Diane* was secured to an inside berth on the Portsmouth State Fish Pier. The boat needed to be right against the pier so the welders who were installing the net reel, winches, and hydraulics could load this heavy equipment aboard, and so their electric leads from their welding machine could reach the boat. The net reel had been prefabricated at their shop based on measurements provided from Hall Quarry and only needed to be welded to the steel plates already installed. This was completed in less than half a day. At that point, I was informed by the owner of the welding company he would need payment for work to date. Further, I would need to provide payment for the winches, which were arriving C.O.D. This was not the contracted arrangement and the bank refused to release the money without a discussion with the owner. Those discussions revealed the company had lost considerable money building an off-shore lobster boat on a fixed-price contract and was nearly broke. They could not pay their workers without the money for the reel and had insufficient funds to cover delivery of the winches and towing wire. The bank negotiated a solution after receiving an itemized bill for the net reel and wrote a bank check to the winch company, thus making sure the winches were actually paid for. The workers here were very different from those at the boatyard. While they had the requisite skills, they needed constant supervision as they were prone to liquid lunches followed by relaxing hits on their bongs. Any work done after lunch had to be checked, especially the hydraulics plumbing. After two weeks, the job was finished, and everything appeared to be in working order. I rode down to Hampton to load the nets and dragging doors.

I was met at the Hampton State Pier by Red and his brother-in-law, Ernie. Ernie had agreed to be my crew until the dragger he was having built in Nova Scotia was completed. This was very helpful, as he had dragging experience and knew the nuances of rigging the gear and checking to make sure it was operating correctly. Dragging is the most technical of the major New England gear types. You need knowledge of geometry and algebra to take measurements to make sure the gear is performing optimally. The net builder, Dan, had ordered the doors based on his fishing experience, but this would still not guarantee the doors were matched correctly to the net. Two boats towing identical nets and doors with identical horsepower would always get different results. The gear gave you clues about performance once you started fishing, but the clues were subtle. I would refer to this phenomenon later in written papers as reading the tea leaves. By late afternoon, the gear was all rigged and we were ready to fish. By this time, we had a attracted a crowd of onlookers in pickup trucks. I always referred to these people as the parking lot police. They were fishermen who formed knots of trucks to gossip, complain, and handicap various events in the harbor. They seemed to spend more time bitching than fishing and wondered why they were always insolvent. Today's prevailing wisdom was that I would be broke in six months, as no one could possibly pay for a boat like this.

The crowd may not have been that far off. After paying for the last of the gear and filling the fuel tanks, Ellen and I had a balance of twenty-four dollars in our accounts. We had a house payment in two weeks and a boat payment in three weeks. I needed to catch fish, lots of fish, and to do it quickly. Ernie and

It takes a community to outfit a boat! *Left to right,* Capt. Ernie Knowles, Capt. David Goethel, and Capt. Marvin "Red" Perkins outfitting the *Ellen Diane* for the first time. Hampton Harbor, May 1983.

I agreed to depart Hampton at 4:30 the next morning. As we were preparing to depart, my father showed up sporting thermoses and a box of donuts. I had not expected him, and I was not sure how Ernie would respond. My father had a way of getting in the way, and I already had my hands full on my first day. Ernie liked to talk and so did my father, and they seemed to get along. I told Ernie to give him orders so he did not get in his way. I figured he would listen to a stranger better than me.

Ernie's style of towing was to set out as close to the harbor as was legally possible, tow three hours to the east, and haul up the net. I let Ernie make the first tow for two reasons: first, because he had the knowledge, and second, because I wanted to let him know how important he was to our success. The gear went out without a hitch, and after several throttle adjustments our speed over the bottom was deemed optimal. Ernie was impressed with the boat's power and how easily the net towed. We made some measurements, and I did some math. From what we could measure on deck, the spread of the doors was optimal, but we would have to wait for haul back to check the bottoms of the trawl doors (called shoes) and the shine on the chain on the bottom of the net to see if the net and doors were doing their best.

By now, Ernie was engrossed in tuning my electronics, several of which he had never seen. I had a color fish machine with a picture that moved continuously across the screen. I also had a side-scan sonar, an outgrowth of World War II submarine technology. This machine had a sound dome that lowered through the bottom of the hull in a watertight tube. The dome emitted pulses that the operator could adjust based on vertical depth and range to target. If the pulses hit something hard, like rocks, the machine would light up in that direction and emit a ping just like in the World War II movies. The function was to allow the fisherman to get near bad bottom without getting too close. Ernie described towing on the *Ellen Diane* as being like operating a spaceship. Watching other boats in the area, I realized Ernie did not tow like everyone else. The Gloucester fleet seemed to be working a constant-depth contour or fathom curve, and Red was making loops around several pieces of bad bottom about five miles offshore. My first thought was, if we caught fish how would we know where? You can see cod and haddock on the fish machine, but flounders are spread out across the bottom, and you cannot mark them with the machine.

At the time Ernie deemed necessary, we engaged the hydraulics, hauled up the doors, disconnected the ground cables from the doors and wound the cables, followed by the net, onto the reel. When the end of the net, called the codend, came up to the stern, the reel stalled. The fish were almost up to the heavy rope called the splitting strap that was wrapped around the net about six feet above the end. We lowered a lifting block down from the rigging, hooked the strap, and carefully lifted the bag of fish over the stern. Ernie pulled

How many captains does it take to wind a net on a net reel? Six captains, including (*second from right*) Capt. Frank Goss, Red, and David. Hampton State Pier, May 1983.

the tripper and the fish rushed out, filling the bin. I estimated we had fifteen hundred pounds of fish on board! Ernie and I checked the wear marks on the shoes of the doors and the shine on the chain sweep of the net, and all appeared to be in order. This time, when we set the net, I ran the gear out and made the tow, while Ernie sorted and boxed the fish. I called Red and discussed the catch and the gear. He said he believed everything was working fine and congratulated us. When Ernie was done boxing the fish, we had about eight hundred pounds of flounder and two hundred pounds of cod and haddock. The second tow produced similar results. When I began to reset the gear, Ernie asked me what I was doing. I told him and his demeanor immediately changed. He had always made two tows and been home for an afternoon nap. I would be making three and be home for dinner. He muttered something about not having enough food. My father plied him with a couple donuts which seemed to stifle that argument.

By the end of the day, we had run out of plastic fish boxes. I had bought thirty boxes from my fish dealer, Tri-Coastal Seafood Coop, and they were all full. We dumped the cod in a fish checker so all the flounders would be boxed. Unloading was now done with electric hoists at the newly constructed Hampton State Pier. Tri-Coastal had a truck with a scale and ice. The totes were plastic and held one hundred pounds, but otherwise this was the same unloading regimen we had done in the past. The parking lot police were in full attendance, even Frank was on the pier and Red had waited to see us. I climbed up in the parking lot to run the hoist and talk with Red and Frank. I could see Frank doing a calculation of the total number of boxes. He summed up the day succinctly with a statement that my gear must be working well and asking how Ernie was doing? Talking with Red, I learned he had done better on cod but had less flounder. Frank went over to have a few words with the parking lot police. I am sure whatever he told them was not what they wanted to hear, as they all soon disappeared. Our total for the day was a little less than thirty-seven hundred pounds. The next day, we would tow differently to try and catch more cod. Cod were worth more than a dollar and flounder averaged fifty cents. In commercial fishing, you fish for dollars, not pounds.

The next several days, I changed our towing to areas near where I had gillnetted. I thought I knew where all the hard bottom was, but soon learned there were hundreds of tiny rock outcroppings that could wreak havoc with a flounder net. Thank goodness for the scanner. Our catches reduced some in pounds but increased in dollars. Ernie was not happy, as he thought towing near the hard bottom would catch rocks and cause problems, but my father kept him plied with donuts. If the donut thing kept up, we would all weigh three hundred pounds by the end of the summer!

At the end of every week, the boat received a check for fish landed. The weeks ran Sunday to Sunday, and you were paid on Friday through the previous Sunday. Now, fishermen do not get paid the same way as people on land. You receive a share of the value of the catch, not a set salary. If a set salary was paid, crew would soon only work in the months where weather made for little work, and you could never get anyone for the summer months when you worked every day. The Supreme Court even weighed in on this issue when the Internal Revenue Service sued New Bedford boat owners over not withholding taxes. The court ruled fishermen were independent businesses paid a share of the catch and thus responsible for their own taxes. How the catch is split was determined years ago. Fuel and ice and food, if necessary, come off the gross, the boat receives 50 percent of the remainder and the captain and crew split the remaining 50 percent based on the number of crew. In my case, with one crew, I received 30 percent and Ernie received 20. The expense of getting the fish to market is deducted by the coop before the boat is paid. Tri-Coastal took seven cents per pound and 7 percent of the value of the fish. So, if the fish was worth one dollar per pound, the coop received fourteen cents from the sale of that pound of fish for their services and expenses. If the *Ellen Diane* received $5,000 for a week's catch, and fuel was $400 and ice $100, the crewman would receive 20 percent of $4,500 and the captain 30, and the boat 50 percent to pay bills such as mortgage, insurance, maintenance, fishing gear, etc. In this example, the boat would receive $2,250, the captain $1,350, and the crewman $900. Ernie's attitude brightened considerably after the first check, when it became obvious the higher catch rate combined with longer days equaled substantially larger paydays for everyone.

One thing I have noticed in all the fisheries, is just when life seems to be going well, disaster is near at hand. In our case, near the end of the second week, the LORAN went haywire. A technician from the electronics company arrived, said the unit had to be returned to the factory for warranty work which would take about four weeks, and if I needed one, I could buy another. To say I was not pleased was a vast understatement. This unit was over three thousand dollars, and I did not have that amount of money. No offer of credit was extended. I called a person in the marine hardware business that I had done business with on the party boats and asked if he had a lesser-priced unit I could buy. Yes, he did, and he would install it and not bill me for thirty days. If I did not have the money by then, he would extend credit further. The attitudes between the two businesses could not have been more different. That shop received every bit of my electronics and marine hardware business from that day forward.

We were back on the water the next day. Towing on nice days weatherwise is straightforward. You can turn pretty much at will and maneuver the net in and out of tight spots. On days with a lot of wind and current, you need a different approach. Turning against the wind may be impossible or take a great deal of ground to execute. Usually, you can hasten the process by turning the steering all the way over in the desired direction and using more power. However, this puts tremendous strain on the rudder, rudder stuffing box, and hydraulic steering rams. My steering in the *Ellen Diane* was sized for a forty-five-foot work boat. My experience with engineers has always been that their calculations are based on ideal conditions, and they usually underestimate real-world forces. Sure enough, one windy day I came out of one of these turns with almost no ability to steer in one direction. I opened the lazarette hatch to find one of the two steering rams had a kink just beyond the halfway point in the steering ram. The ram could not retract beyond the kink into the piston. We had to haul back and limp home, being careful to make sure all turns could be executed in one direction. Safely tied to the pier, I called Hall Quarry to explain the problem. I was discouraged as this was not a matter of replacing a defective piece of hardware. We would need an entirely different steering system that would require massive reconstruction in the lazarette. The yard owner was amazed, saying he used this same unit in all his boats and never had a problem. The difference was, and we had this discussion numerous times during construction, the forces placed on a vessel in dragging are substantially higher than gillnetting or lobstering. The steering manufacturer was in Washington State, so despite the late hour, it was still open. An hour later the phone rang. A new steering ram had been located at a marine hardware store in Portland, Maine. This was sized for a ninety-foot work boat and the manufacturer was certain I could not bend it. The steering mount would have to be moved and fiber-glassed into the lazarette, as my unit was a dual-piston model with short rams and this was a single-ram model with a long throw.

The owner would work with Charlie that night, using the boat currently under construction as a template, to make a blueprint for Charlie to follow when he arrived in Hampton. Charlie would leave Hall Quarry at an hour appropriate to place him at the marine hardware store when it opened. With luck, he should be in Hampton by eleven, and the owner thought he could complete the installation late that night or early the next morning. I was stunned, once again someone who wanted to see me succeed was going way above and beyond on my behalf. Charlie arrived early and went straight to work. His workspace was cramped, dirty, poorly lit, and the conditions were stiflingly hot, but by nine that night a functioning steering system was in place. Charlie had installed many overlapping sheets of fiberglass, and although it was already hard to the touch, he wished to let it set up overnight before trying it out. Charlie came back to my house, Ellen provided us with a good hot meal, and Charlie turned in for some well-earned rest. Early the next morning, we did some tests of the steering throw and checked for flex in the new glass. Finding all acceptable, we went for a ride and put the new system through its paces, using backing down with full reverse with the rudder hard over as a proxy for making turns while towing. Everything seemed fine, so I returned Charlie to the pier, thanked him profusely, and Ernie and I headed out to fish.

Two times in a single month, people who could easily have looked the other way and put me in a very difficult, if not impossible, financial bind stepped forward to keep me fishing. I tried to pay it forward. The marine hardware store was fully paid for the LORAN by mid-June, both house- and boat-mortgage payments were made on time, and the boatyard was fully paid for the boat construction overrun by mid-September. I inquired several times about charges for the new steering. I was told the yard and steering manufacturer had reached an agreement and I bore no financial responsibility.

Ten Thousand Years Too Late and Baby Duty

Dragging settled into a predictable routine. The flounder catch began to decline by the end of June, but there were still cod around the myriad humps that dot Ipswich Bay. Most of the local boats quit fishing and prepared to go tuna fishing for the summer and fall, but Red and I kept at ground fishing. I would often encounter many Gloucester boats fishing for silver hake. These boats were generally larger and rigged differently than the local boats. Their form of rigging was called an eastern rig, and they towed their nets from two frames on the starboard side of their boats. One frame was near the bow and the other near the stern where the wheelhouse was located. They did not have net reels and had to strap their nets in bights, which they lifted in with blocks rigged from a tall mast. This setup required more men to operate, and there were usually at least four men per vessel, often blood relatives. Because they towed from the starboard side, they could turn very easily to starboard but had great difficulty turning to port.

We towed from the stern, which allowed us to turn either direction. This made it possible to encircle humps, whereas the eastern rigs only approached bottom on their port side so they could turn quickly away to starboard. This meant captain Fred's first rule of the road was often in play: the big march on and the little sink. One day, I encountered a boat named the *Sea Fox*. The captain answered the radio, said his name was Tom, and asked why I was turning all over the place. I told him I was edging the humps for cod. He was whiting fishing and wanted to follow the fathom curve through the humps. I gave way, he thanked me, we talked for a while, and he offered me his home phone number with an invitation to meet in Gloucester. We were similar in age and ambition. His fleet nickname was Magellan because he would drive all over the ocean trying differ-ent places for fish. It also turned out he was related to Sal, one of the men I had worked with at the aquarium. This was the beginning of a lifelong friendship.

Most of the fishermen worked together, even though in essence we were competing for fish, but there are always a couple that need to have their attitude adjusted. One early summer morning, I was towing up the edge of the bottom southwest of the Isles of Shoals known as Whaleback. I was towing in and out, using the scanner, when an unfamiliar eastern rig approached dangerously close to my starboard side. If he did not turn, we would either collide or I would hang up on the bottom. I called on the radio and got no response, so I called Tom and asked him to call the guy. He got no response either and told me the boat was more rot than wood, and if I turned sharply at him, he would move. By now we were close enough I could see the captain in the wheelhouse grinning, smoking a fat cigar. I advanced the throttle, turned the wheel hard to starboard, and aimed for his midship. He stopped smirking in short order, the wheelhouse door flew open, the cigar fell out of his mouth, then he dove back inside. I was now less than one hundred feet from his side, moving at about five knots. Black smoke belched from the offender's exhaust, and he took off to starboard. I probably came within fifty feet of his stern, and his radio soon had a miraculous recovery. I heard him telling his buddies some crazy guy in a little white boat was trying to cut him in half. He was faster than me, but I made him work for his escape, following for about three miles before breaking off the chase. Tom tossed salt in the wound, telling him I was his friend, publicly shaming him for his actions. No Gloucester boat ever played chicken with the *Ellen Diane* again.

By early fall, Ellen and I already had the boat payments for the winter in the boat account. We decided to have a second child, who was rapidly conceived and due in early May. I fished through the winter, but not without one big scare. December 7, I remember the date because this was Pearl Harbor Day, dawned cloudy and windy out of the southwest. My boat was in the second row of moorings outside the iron wall where the hoists were located. Hampton Harbor has a long distance to the southwest before you encounter land, probably five miles at high tide, so the harbor can build some good-sized waves. Those waves used to crash harmlessly on the shore, but now encountered this vertical steel wall and bounced back against the oncoming waves. A crowd gathered in the parking lot to stand watch as the wind continued to increase, but nothing could really be done. I had waves breaking over the bow on the mooring and I was convinced either the mooring lines would break or the big cement blocks would start dragging across the bottom. My neighbor's boat, being closer to the wall, had a much more serious problem. In addition to waves over the bow, the bounce-back waves were going over its stern. Sure enough, later that afternoon when wind gusts reached ninety-seven miles per hour on a local wind gauge, the boat gradually settled beneath the waves. The owner and captain seemed unusually calm for such a traumatic event. The nearly empty bottle of Jack Daniels sitting in the passenger seat probably was partially responsible. My boat miraculously survived, and the next day the entire harbor turned out to retrieve the sunken dragger so it could be rebuilt.

By now, Ernie's new dragger had arrived and he was fishing on his own. I was going alone and looking for crew. Ellen and Eric were out walking when they encountered the sunken boat's captain's wife. She confided they were in serious financial difficulty and needed money to buy heating oil. She wondered if her husband, Merrill, might be able to make a few trips with me. Merrill was a character, often described by locals as having been born ten thousand years too late. He frequented the bars and had serious anger-management issues, as many young wiseasses found out the hard way. He was the only man I knew of that the cops would offer a ride home at last call, as it was much safer for their wellbeing than trying to slap cuffs on him. We reached an agreement that we would fish together, but if he wanted to go to a local watering hole, I would leave him there and he could find his way home. Being winter, we did not get out often, but Merrill knew some winter cod spots that I had never fished, and we had some big tows. Being winter, we had to keep our skiffs in our trucks to keep the harbor ice from sinking them. One night after dropping Merrill at a local tavern, I was driving home up the Hampton expressway. As I slowed down and began to turn off the highway, unbeknownst to me, the skiff slid out of the truck onto the road. When I reached home about five minutes later, I noticed the missing skiff, but had no idea where I lost it. I retraced my way and found the skiff with a car containing an intoxicated person parked next to it. I turned the truck around and enlisted his help in placing the skiff back in the truck. I could see blue lights coming up the road and told the guy he better beat it, but as I drove off, he was just sitting there muttering about seeing boats in the middle of the road.

Ellen and I took a vacation in February, and I let Merrill run the *Ellen Diane*. I was nervous and called home to check every night. Initially, he was upbeat saying he had found some gray sole east of the Isles of Shoals, but as the week wore on, he became more evasive about what was happening. When we got home, I found he had all three nets laid out in a parking lot in various states of destruction. What was worse, he had broken the capstan trying to lift in a large bag of sole. Judging by the condition of the codend, the large bag of sole was a giant rock with a few fish around it. At any rate, if I would go buy half a bale of dragger twine he could fix everything. To give the devil his due, he repaired all three nets and had some machine-shop buddies fix the capstan. By now, his boat was almost ready for the water and our time together was over.

By mid-May, Ellen was overdue and huge. The doctors had done an ultrasound to check for twins, but there was just one very large, content baby. Occasionally, I would put my ear down to listen for a heartbeat, and the baby would unleash a kick. This was like getting punched in the face. We were informed that, absent birth in a few days, labor would be induced. That conversation must have been heard internally because labor soon began. Ellen had another long, difficult labor, but when this child decided to appear he just came out. The nurse was just finishing wrapping him in the blanket when our pediatrician stopped by on rounds. He took one look at our son, who we had named Daniel, and exclaimed, "Well, take this kid out for a steak and a beer." Daniel was very different from our first son, Eric. His only requirement was food and lots of it. The only time I saw him truly upset in his first two years of life was when he toddled into the kitchen to open the refrigerator and retrieve his sippy cup of milk. The handle snapped off the refrigerator without the door opening, and, with elephant tears streaming down his face, Daniel returned to the living room dragging the handle, exclaiming "Rator broke, rator broke!"

The remainder of the year was relatively uneventful. The manager of Tri-Coastal asked me to consider whiting and herring fishing for tuna bait. I bought a used net and brought it down to Dan for repairs. He was not impressed with my purchase, which had serious rodent damage. He also said it was not hung correctly on the footrope. I figured he was going to tell me to put it in a dumpster, but he was a twine man, and this was a challenge. A week later, I picked up the finished product, which had every color piece of twine imaginable, but it worked. I sold the bait to the coop into the fall, easy duty and short days. Late that winter, the manager again approached me about fishing for northern shrimp. He had some Japanese customers who would pay top dollar and the rest he could sell to a processer. I knew nothing about shrimp; in fact, I did not even know they lived here. I called my friend Tom, and he explained that the shrimp start in eighty to ninety fathoms in December and move into the shoal water, sometimes only twenty-five to thirty fathoms, to spawn in late January and February before retreating offshore again in the spring. My results were mixed. Tom caught seriously more shrimp than me, and it was obvious I would need to have a new, modern net built. Furthermore, in the deep water, the net was dirty catching small flounders and small whiting. A new kind of sweep called a rockhopper was being used by some shrimp boats with very good results, both on shrimp catch and bycatch reduction. I decided to call a net builder who was advertising these new sweeps and have a net built by the next shrimp season.

By the spring of 1985, the ocean was getting crowded. Every harbor had more boats, and cumulatively they were adding up. Towing was beginning to look like what British mariners must have seen at Dunkirk. Everyone was turning, arguing just trying to get by. In the middle of May, while trying to avoid a knot of boats of the end of a large hump, I made an extreme turn to escape. When you make a turn like this, the net is traveling over different bottom than

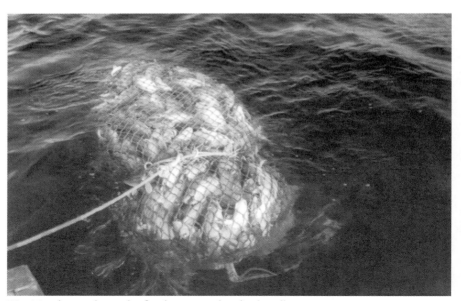

Wow! Six thousand pounds of cod is too much to haul in all at once. May 1984.

the boat. In this case ,the doors whacked a small hump I did not even know was there. I quickly turned back the other way and waited for the boat to stop, but there were a couple tugs and I kept going. I turned off several other humps and noticed the boat was not turning easily. Thinking I might have picked up something, I decided to haul the net up. After the doors were up, I noticed the net was floating right on the surface

No room on deck! David and a load of large cod escaping the fish bin, 1985.

and several gulls were standing on the codend. As the reel continued winding, I realized the codend was completely full of big cod which were floating on the surface. We had to split the bag four times to get all the fish in the boat and when we were done, the entire deck was covered with twenty to seventy-pound cod. I went home as it would take a couple hours to clean all these fish. When we were done, I had unloaded a shade less than six thousand pounds of cod.

Red and I had purchased radios with unassigned channels in hopes of insuring private conversations. I told him to call me when he got home. I knew which three humps I had grazed but was not sure if one or all three had fish. We hatched a plan to set out some distance from the area and then head off in diametrically different directions. We both set out with probably fifteen boats within a mile of us. The gear hit the bottom, and I turned south while he turned north. Pandemonium ensued as everybody scattered. Two boats hooked their gear together, others had to stop; it was wild, but hopefully it served a purpose: you cannot catch what has already been caught. On the first tow, I went by the hump I judged least likely to have had the fish. I was right and only got one hundred pounds of cod. I repeated the process with the second tow and the second hump. That meant all the fish had come off the hump I had banged the doors on.

Fish are my friends! David and a one-hundred-twenty-pound cod. June 1985. Photo by Jack Goethel.

I communicated all this information to Red, who had been busy checking humps in similar depths with similar results. At this point, the wind picked up from the south and the fog shut in. Now no one knew which boat was who. I lined the hump up and laid one door on its edge. I could clearly see a big mark of cod on the edge. I towed past the hump about a mile and hauled back. I told Red to line it up. I had about thirty-four hundred pounds and Red had close to two thousand pounds. We went home and unloaded. Some people really gave me a hard time, but I had seen boats beat on pieces of bottom until they caught every fish. I thought this was short-sighted, and since I planned to do this for a long time, I was thinking about sustainability on a longer time scale. The rest of the people fishing never found those fish and after several weeks the school moved on.

CHAPTER 17

How to Catch a Fish

The cod catch gradually dwindled around that hump, but we had done extraordinarily well. I used some of the proceeds to further pay down my boat mortgage and the rest to buy a new, small-mesh net. I drove to Connecticut to speak with the owner of a net shop who had some new ideas on how to build nets to catch small-mesh species. Peter had studied under a West Coast net builder, who himself had gone to Europe to investigate innovative ideas. From what I had read, Europe was probably twenty years ahead of the United States in both net and trawl-door design. Fuel cost much more in Europe, so any net and door combination that had less drag and more catching efficiency was fully exploited. Peter proposed to build a four-seam box net which would be attached to an eight-inch rockhopper sweep. The net would look like a tapered box as it was towed through the water instead of an elongated funnel, which was the shape of local nets. The rockhopper, which was made of round rubber discs punched out of old tires strung very tightly on a wire, would keep the net off the bottom and jump over small rocks. The net would not disturb species like flounders, skates, monkfish, and lobsters, which would pass underneath the twine suspended above the rubber discs. The net would also reduce the catch of small whiting, which appeared to charge the space between the discs to escape. This would save the crew a lot of monotonous work during shrimp season. Peter showed me net plans and pictures taken in test tanks of mockup nets to illustrate his points. I decided to place an order to have a net constructed before returning to New Hampshire.

As had become the pattern, the cod magically left Ipswich Bay by mid-July, and the whiting, herring, and tuna began to show up. Tri-Coastal again wanted bait. The preferred bait was herring, so I tried to target these fish. Herring is considered a pelagic species; that is, they swim primarily well off of the bottom, but we had noticed that most mornings, right at sunrise, they spent several hours on or near the bottom. My new net opened more than traditional two-seam nets and caught these herring that were eight to twelve feet off the bottom. As advertised, it also caught far fewer bottom-tending species that boats were not allowed to keep when using small mesh, making the job of the crewman much easier. Herring forms very dense schools, so when you catch them, you catch a lot. I had fifteen-minute tows with over ten thousand pounds. Now I had a new problem. The coop could use about fifty boxes a day. When I had big days, I started selling the surplus to lobster boats in Hampton Harbor and to several people in Portsmouth. I had a friend who lived in Rye who wanted big mud hake. The coop tuna boats really did not want them because they were tough to cut up for chum. So, everything was used, which is the way I believe fisheries should be run.

I spent large amounts of time unloading fish at the pier in Hampton. We often attracted a crowd, as people are curious to see what you have caught. Inevitably people would ask, "How do you catch a fish?" I could write a book on this subject, but basically it boils down to knowing three items about your intended target. Fish are very basic organisms. Bottom-dwelling fish seek to maintain a suitable bottom-water temperature. That temperature, in turn, attracts prey species for the fish to eat. The fish are driven by a primal urge to spawn at a certain time in a certain temperature and usually return to their spawning grounds every year. If you know these three things, you should be successful at catching fish. Fishery scientists often overthink this; food, bottom-water temperature, and sex determine fish distribution. Now, to catch particular species of fish you need a net that is built to target that fish. As much as fishermen would like to have a net that catches everything in every place, such a net does not exist as of this writing. For example, if you wish to target the four types of flounder that live in the Gulf of Maine, you need a net with either a chain sweep or one with very small rubber discs. Flounders lay on the bottom and the sweep must flip them over into the bottom of the net. Loops of chain are the most efficient but come with a cost. They will roll a rock over into the net. Discs are a little more forgiving as they will jump over rocks half buried in the mud, but also miss a few flounders. If you want to catch haddock, you can use the same two types of nets, as haddock also swim over the mud, but you need more height because haddock swim off the bottom. This means you need to change the configuration of the net to get the top higher in the water column. Haddock are very responsive to small changes in bottom-water temperature, so being able to measure the temperature is important. I have often seen a half-degree difference in temperature change the haddock catch from one hundred pounds per tow to one thousand pounds. The direction of the wind and phase of the moon also radically alter haddock catch rates.

Cod to me are the most interesting fish to catch. They occur in different places at different times of the year and require different strategies. You can use a flounder net in the spring to catch cod. At that time, they are found along the edges of deep-water humps surrounded by mud or sand. Later into the summer, cod will move away from the humps just at daylight to feed, but once the sun is up in the sky, they retreat to small, rocky outcroppings where they hover over the bottom for the remainder of the day. To catch these cod, you need a rockhopper sweep with a lot of flotation so it will jump when you encounter the ledge. Rockhopper fishing is high risk/high reward. If you make it over a hump successfully, you can get a week's pay in twenty minutes. If you hang the net on the bottom, you can do thousands of dollars in damage and spend the week repairing the net. The most successful way to do this is to ride back and forth over the hump before setting the net. Humps usually have a steep side and a side with a gradual rise. You want to tow up the steep edge so when the gear jumps off the bottom the twine does not land on the peak and then down the more gradual edge. Get far enough from the peak so you can set the gear and towing wire and have enough room for the net to open and the doors achieve their correct geometry. Use less towing wire than when fishing for other species, so that when the doors encounter the ledge, they will be more likely to fly off the bottom. Increase the speed as much as possible, as the boat will slow down when the net and doors encounter boulders. Finally, do not turn on the bottom. Turning causes the net to change shape and lays twine down on the bottom. If the boat stops, haul back and hope you have enough net left to try again. If you make it, tow the net for twenty to thirty minutes to tire out the cod, so they do not swim out of the net during haul back. Always keep the boat moving forward until the codend is against the stern. Often, the net will have big rips, and if the fish can swim forward, they will exit through the holes. Hopefully, you will have a codend full of cod and spend a few hours cleaning fish. Just as often, you will have very little and spend many hours fixing torn twine.

In the late fall, the cod move up on the edge of Jeffreys Ledge to feed on herring and herring spawn. Here again, you need a rockhopper, as the bottom is mostly sand with a few small rocks. The area you fish is a narrow band between twenty-eight and thirty-two fathoms. Shallower than twenty-eight fathoms and the bottom is untowable ledge; deeper than thirty-two fathoms, the bottom changes to mud and holds no cod. Usually, you tow about an hour. These cod will be smaller in size and full of herring. Catch rates range from a few hundred pounds to a few thousand pounds. When you reach the dead of winter, mid-January to mid-March, look for cod in the very deepest parts of the deep-water basins. These are often subtle indentations in the bottom three to six feet deeper than the surrounding mud. What I have found is the water here is slightly warmer than the surrounding mud and fish form very dense schools, probably to stay warm. Perhaps there is some deep-water spring in these places. At any rate, I have had enormous catches of cod when I locate these holes, sometimes five to ten thousand pounds, while a boat five hundred feet away catches three fish.

All fish have their quirks. Monkfish move during full moon. Pollack touch down on the bottom during the change of the tide. If herring are going to touch down on the bottom, they do so right at sunrise. Shrimp, although not a fish, move against the direction of the wind. When the wind is from the east, the shrimp head east for deep water, conversely any westerly wind moves them into shallow water near shore. Of course, all these pointers are generalities. If fish were this predicable, it would be called catching, not fishing! I also think it is worth noting that scientists and fishermen are very similar in their problem solving. I can hear elements in both groups now saying no way, but consider, both create a hypothesis and then test it. Both keep very careful observations so they can understand the natural world that can only be inferred. Both test their hypothesis numerous times to make sure there is

A great day's pay and a lobster dinner! David, with thirty-five hundred pounds of cod and an eighteen-pound lobster. Photo by Jack Goethel, 1986.

repeatability in the results. I always considered every day at sea to be an experiment. All the successful fishermen I know approach fishing in this manner.

Strip Miners

One day in early fall, I went to visit my friend Erik. I had finished up early, having delivered a boatload of herring and been home for lunch. Erik had taken a job at the moving company and was working as a dispatcher in Portsmouth. I walked into the office and found Erik with one cigarette in his ashtray, one in his mouth, and an unlit one behind one ear and a pencil behind the other. He was talking on one phone while putting someone on hold on another. Erik looked like he had not seen the sun in months, and, quite frankly, I was shocked. His job apparently was to straighten out the daily snafus that occurred between the moving crews, customers, government representatives, and anyone else who had something to bitch about. After about half an hour, he was down to one cigarette and no one on the phone. I told him bluntly he looked like hell and ought to start looking for a boat. I was quite sure with his fishing skills he would do fine.

The pep talk must have had some effect because about two months later he called to tell me he had purchased a gillnet boat. Even though we were using different gears and fished many times in different places, we swapped information and worked together for the remainder of our careers. Fishing was what we both wanted to do, but fishery management was what we both ended up doing. Erik drew the short straw and was appointed to the New England Fishery Management Council in the coming decade. The United States had taken Canada to court over who could fish on Georges Bank and lost at the international court in the Hague in the fall of 1984. The waters where the majority of the groundfish were located were now Canadian territory. The grounds already had too many boats chasing too few fish because of the government-inspired building boom of the late 1970s. This fleet now packed into very crowded waters which were soon decimated. The chickens would take a while to come home to roost, but there was no doubt of their impending arrival.

Fishery management was still in its infancy. The structure set up by the Magnuson-Stevens Act created eight regional management councils which wrote management plans that the National Marine Fisheries Service could accept, reject on legal grounds, or reject a specific component of on legal grounds. Prior to federal management, an International Commission for Northwest Atlantic Fisheries had very nominal control, basically setting some quotas it could not enforce and a mesh size that was laughably small. People just went fishing; American boats fished within sight of Canada and vice versa. In New England, 25 percent of the boats caught 75 percent of the fish. An obvious place to begin managing the fisheries would be here. Appointments to the council were political; you were nominated by your state's governor and the Department of Commerce picked the appointees. The fishermen appointments on the council, not surprisingly, went to the large fleet owners and their lobbyists. The council passed regulations at a dizzying rate: minimum fish sizes, minimum mesh sizes, closed areas, gear-marking requirements, small mesh requirements, and a host of other confusing and mostly useless junk.

In 1986, I attended a council meeting and spoke about there being too many large boats chasing a declining number of fish. Fishermen could see this even if science had not caught up with views on the water. The council chair at the time owned a fleet of those boats. I guess I looked young for my age as I was often addressed as "Sonny" by older men who did not know my name. I was dressed down with "Sonny, we are trained professionals around this table, and we know exactly what we are doing. Now why don't you go back where you came from and go fishing and leave management to us." He was right about one thing; they knew what they were doing. After they strip mined Georges Bank and the deep water of the Gulf of Maine, he sent his boats to fish in Alaska.

CHAPTER 18

Annual Meetings

By the winter of 1987, I was being urged to take a more active role in fishery management by the manager of the Tri-Coastal Seafood Cooperative. Tri-Coastal had been formed in the late 1970s to market members' bluefin tuna. Cooperatives are different than most American businesses in that members are allowed to band together to get better prices for their products. They can set prices or sell members' products for the highest price offered. In any other business, this type of marketing would violate United States anti-trust laws. Cooperatives are corporations or associations organized under the Fishery Cooperative Act of 1934, a piece of depression era legislation designed to help struggling fishermen. Most cooperatives are small and oriented to serve a geographic region, but some have grown to national or even international representation. Readers may be familiar with Land O'Lakes, a dairy cooperative headquartered in Minnesota, or Ocean Spray, a cranberry-grower cooperative headquartered on Cape Cod. Each member buys one share in the coop, thus everyone is equal, and should the cooperative make a profit, each member gets a performance bonus dividend based on their contributed sales. Tri-Coastal had been formed to market bluefin to Japanese buyers, which had increased the value of the fish dramatically. Tuna, however, was a seasonal business, June to October, and the manager realized he needed other types of fish landings to keep the business viable. They began buying groundfish in Newburyport and Hampton in the late 1970s. This offered fishermen like Red and me an opportunity to have someone sell our fish and perhaps return some extra money at the end of the year. This was very different from selling fish to a for-profit buyer who paid you what he felt like paying if he paid you at all. Fish dealers were notorious in this period. I questioned a buyer once about the difference between what I received and what I was promised and had a German shepherd with a bad attitude turned loose on me to discourage further collection efforts. Gloucester was especially bad, as buyers would meet boats with bottles of whiskey and encourage consumption while the captain was trying to watch the scale. For the cost of a fifth, the buyer could short the boat thousands of pounds of fish. So, coops in both lobsters and fish were formed up and down the coast of New England to combat the rampant corruption of fish dealers.

Other members of the coop asked me to run for the Tri-Coastal board of directors because I was both a major groundfish producer and because they felt my knowledge of biology would help us fight onerous regulations being proposed for tuna. The government had started using a new fish-assessment model called a virtual population analysis or VPA, which is a very complicated mathematical model relying extensively on correctly aging the fish samples. The government at this time also supported a two-stock model for tuna which had the fish in the western Atlantic biologically distinct from the eastern Atlantic. This theory proved very convenient for the State Department, which had quietly negotiated military base rights in southern European countries in return for turning a blind eye on rampant overfishing of tuna in the region. The first presentation of this new information was done in Boston. A government scientist named Mather told the stunned crowd of fishermen that there were only four thousand adult tunas left in the western Atlantic and fishing must cease. The crowd was particularly irate because the combined catch of Canadian, American, and Japanese fishermen who fished the high-seas, Mid-Atlantic Ridge in the central Atlantic exceeded four thousand fish! This was the beginning of New England's deep distrust of fishery science, which persists to this day. Fishermen demanded a do-over and further demanded examination of the two-stock theory, which they did not believe. If bluefin tuna fishing in the western Atlantic was going to be virtually shut down, they wanted a shutdown of the European fishery because they believed fish traveled back

Superman (four-year-old Eric Goethel). We need a bigger boat! *Left to right*, Ted Fyrberg (Tri-Coastal Seafood Coop), Eric, and David. Gloucester Harbor, 1985.

and forth across the Atlantic. This concept would be proved by tagging studies twenty years later and the concept that fish were political pawns would be proved by Carmel Finley's *All the Fish in the Sea* in 2011 using newly declassified government documents. The concept of maximum sustainable yield was created by the State Department in the 1950s as a geo-political strategy disguised as science to make sure some countries fished and others did not.

Our manager, perhaps being more pragmatic than emotional, took a different

Our future fisheries scientist, Daniel Goethel, up close and personal with his dad's giant bluefin, helping load onto the truck. Hampton State Pier, NH, 1988.

approach. Tri-Coastal filed a Freedom of Information (FOIA) request for all the documentation of this new science and had it shipped to my house. Between Christmas and New Year's, a five-foot-high pile of boxes full of paper arrived. I warned everybody I was a research biologist, not a mathematician, but still spent the winter on my days off sorting and reading this mound of paper. I learned the importance of correctly aging the fish, but had no way to independently verify the results, and I also found some intriguing documents marked confidential where scientists warned the State Department that significant intermixing occurred between the two stocks of tuna and European catch rates would have to be lowered. Ultimately, the Japanese longline fleet would abandon the western Atlantic and American and Canadian fishermen would receive small quotas for the foreseeable future. About five years later, the same scientist who said there were only four thousand fish left found large mistakes in the aging of fish between nine and thirteen years old. Rerunning the math produced much higher fish estimates, but the damage both to fishermen and science endured.

During this same period, a new set of players entered the fishery management arena. Environmental non-governmental organizations (ENGOs) started showing up at council meetings. Initially, they were international groups such as World Wildlife Fund and Greenpeace, known mostly for their headline-grabbing stunts, but soon a whole host of home-grown groups joined the fray. Their modus operandi was to select an animal, preferably with two eyes that faced forward, and start a campaign to save it. The goal was making money and lots of it: mail out millions of postcards, hopefully tugging on the heartstrings of housewives across the country and get them to mail back donations. Flush with cash, they would then hold demonstrations and intimidation campaigns. If that did not get them what they wanted, an army of lawyers would be sent in to sue the government, and, of course, if they won, terms of the settlement would pay their legal fees. To my mind, this was nothing more than legalized extortion and fishermen were to be collateral damage. To make matters worse, they often had Hollywood celebrities as major donors and spokesmen. I have always been amazed at how many people in this country take their lead from some man or woman blessed with a pretty face, but who is otherwise completely incapable of intelligent conversation in their chosen field. The ENGO Oceana led off every statement with something about Ted Danson, like anyone listening cared. A woman representing the Conservation Law Foundation once told a fishing acquaintance within ear shot that her views on the record were nothing personal, just your job or mine. At least she got an A for honesty. The problem was, at least with groundfish, these groups had a point. There were too many boats chasing a fixed and declining number of fish. This was the result of several factors. The government tax credit to build new, modern boats combined with the loss of the eastern Georges Bank—which is where the largest volume of fish was caught—to the Canadians, guaranteed the United States groundfish fleet was well over-capacity for the grounds it could fish. For several more years, the dedicated groundfish fleet would keep regulators regulating the people behind the tree, such as small-mesh fisheries like shrimp and silver hake, but finally there was nowhere else to lay the blame. Two amendments in rapid succession would forever alter the historic New England fisheries. Groundfish Amendment 5 limited the amount of groundfish permits and Amendment 7 limited the number of days fishermen could fish. Each of these amendments had major shortcomings.

To get a permit under Amendment 5, you only had to produce a receipt for one pound of fish sold. While most vessels were limited to eighty-eight days of fishing in Amendment 7 those who would swear to fishing as much as 350 days would receive half that number or 175 days. Thus, the people who had caught most of the fish took a far smaller cut than the majority, who in turn would retain far fewer days to target groundfish. Since there were no poundage controls yet, days at sea served as an indirect way to control catch size.

Against this backdrop, day-to-day life carried on. Red and I bought voice scramblers for our radios so we could discuss cod hot spots. Likewise, Tom and I bought them for discussing whiting catch in the summer. The ocean was crowded, and this was survival of the fittest. We were all still making a living but had to be more circumspect in discussing our catches openly. In 1989, I was elected president of Tri-Coastal Cooperative by the board of directors. This would be a formidable task, as it is difficult to have a successful fish coop when you cannot land fish. Tuna, which is what Tri-Coastal had been set up for, was severely restricted and now groundfish, which Tri-Coastal had used as a back-stop, was being severely restricted. By now, I was attending more and more council meetings, and was not terribly impressed by what I witnessed. I did not like public speaking, but I also did not like what I heard from the ENGOs or some of the fishing groups representing the full-time groundfish boats. The ENGOs wanted huge, year-round closed areas without respect to whether small boats could get around them, and the large boats wanted restrictions on everything imaginable as long as there was no impact on their ability to fish groundfish year-round. Frustrated, I went to the microphone to suggest no fishing at night to let the cod spawn, doing a study to figure out how many boats could fish by size, class, and gear type, and restricting the largest vessels to the offshore areas where you need vessels of that size to safely fish. Then scrap excess capacity by size class to be paid for with a government buyback. I managed to do in two minutes what no one else had done in fifteen years: the entire room was united in hatred of my idea. The ENGOs were furious that areas might be closed only at night. The big boats were furious they might be restricted to offshore areas, and no fisherman liked the idea that they might be forced to scrap their boats. I can honestly say being the skunk at the garden party was a high-risk adventure. I did, however, get the attention of a man from the national headquarters of

National Marine Fisheries Service (NMFS). He culled me out of the crowd in the hall, made sure no one could hear, and told me I could obviously find a job on land and should get out of the fishery now. He said NMFS had a plan that he would deny if I ever went to the press: twenty years from then, zero to fifty miles would be recreational only, and beyond fifty miles there would be a small fleet of factory trawlers catching the commercial quota. At the time, I dismissed his comments as scare tactics, but considering how fishery management currently stands, I would now call him the only honest man in government. Thirty years later, in Groundfish Amendment 18, it is theoretically possible for seven or eight corporations to own all the commercial quota, and most fishing grounds within twenty miles of shore are either permanently or seasonally closed to commercial fishing, though recreational fishing is still allowed.

Against this background of constant turmoil, fishermen had to struggle to maintain a home life. Fishermen's wives were always pulling double duty, raising kids while also running the family business while their boat was out at sea. Fishermen often fish seven days a week when the weather is good and miss out on events in their kids' lives like ball games or scouts, even birthdays. Conversely, when the weather is bad, they may be the only parent in attendance when all the other parents are working. Children of fishermen are far more connected to the environment than most of their peers because they learn to live with the consequences. Similarly, fishermen's wives are better at muti-tasking than many other women because it is what they do daily. Ellen procured supplies for me, made sure everyone got to their appointed place at the appointed hour, learned to represent me at meetings when I was at sea, participated in town government, and started her own business all before our oldest was ten. Her story is not unique; you can find women doing the same thing in every harbor up and down the coast. Behind every successful fishing boat, is a strong, disciplined, no-nonsense, hard-working woman.

In the spring of 1989, I convened my first annual meeting of Tri-Coastal Seafood Coop as president. Annual meetings are required every year to present a financial overview of the previous year, elect the board of directors, ratify any changes to the by-laws, and discuss new business. Apparently, the custom had been to buy attendees dinner at a local restaurant to insure a quorum. Managing a coop is a lot like herding cats. Imagine four corner gas stations competing to sell gas all day, then working together to buy gas for their stations at the best price. Coops work best when they hire a strong manager and then let that person manage. They do not work at all when every board member thinks they are the captain and no one is the crew.

Annual meetings, I soon found out, work best when the strongest thing served is coffee and the business portion is completed before dinner. That would be a prerequisite for all meetings going forward while I was president. However, tonight the restaurant was making a fortune serving alcohol to a group of fishermen rapidly becoming intoxicated. I tried to start the meeting and get dinner served, but hostilities between various members made progress impossible. The manager was a very big man with an equally big personality who had always kept people in line by cracking heads. Finally, he waded into the tables, took a couple of the feuding parties, cracked their heads together, and slammed them into their chairs. The business portion of the meeting was in session. Dinner was being served, and I thought the meeting would soon be over, when someone else stood up, dropped his pants, and told the manager to kiss his ass. This produced another scrum.

At the next board meeting, I threatened to resign if the board did not endorse a no-alcohol policy at annual meetings. Reluctantly, this was agreed to, but the following year we had another incident. A man dressed in a suit coat who looked to be about eighty, keeled over on the floor in the middle of the meeting. I figured he'd had a heart attack and rushed forward to start CPR, but before I got to him, several neighbors had emptied his pockets and knelt over him. They produced two empty flasks and six empty nips with which he had been spiking the coffee I'd had the restaurant serve. He had not had a heart attack; he had passed out drunk! I would be president of Tri-Coastal until 1998, when the coop was liquidated. Most of the membership had grown old and no longer fished. Waterfront property had risen substantially in value. The majority wanted to cash out.

CHAPTER 19

An American Original

My father had fished with me every day since I started dragging in 1983. He had remained sober and quit cigarette smoking, so he was no longer annoying to be around. Some crewmen liked having him on the boat, but a few did not. He meant well but had little sea sense and tended to put his hands where they did not belong. More than once, he got his hands stabbed by flounder picks. On land, ex-wife number three had left him nearly broke. I did not know quite how financially strapped he was until I was requested to cosign a mortgage for a small fixer-up house he bought in Exeter. He had a new lady friend, the town clerk of Hampton. They were a pair of American originals; both divorced several times. She had a side business, a small wedding chapel where she performed marriages. I always thought it humorous that two people who had such poor marriage track records were in the business of performing weddings. She was smart enough to keep her own house, and they never married. At least they both knew their limitations.

My father had beaten two of his addictions, but he still liked to gamble. America always likes a comeback story, and he was on his way back. He began reading various financial journals and started investing in penny stocks. Within five years, he had amassed a nest egg and paid off his mortgage. He also liked the dog track in neighboring Seabrook. One day, he showed up and offered to take our son Eric out to lunch and then spend the afternoon with him. My father had never shown any interest in being a grandfather. Ellen's parents, on the other hand, spent large amounts of time with both our sons. My father showed up for birthdays and holidays and that was about all. Eric was between eight and ten for this first boys' afternoon out. Ellen became worried when by five thirty they had not returned, but soon a pair of headlights pulled in the driveway. Eric burst through the door dragging a huge garbage bag full of something. "Grandpa let me bet two dollars of his money at the track and I won!" he exclaimed. Their afternoon out was spent at the matinee races at the Seabrook dog track. To keep

Eric occupied, so he could study his racing form, Grandpa had procured a trash bag and put Eric to work picking up people's discarded tickets on the floor. He told Eric they would go through them and see if anyone had discarded a winning ticket. I thought my wife would pass out, but she recovered and calmly told Eric, Grandpa would take the bag of tickets home with him, and he needed to go upstairs for a bath. After they were out of earshot, I asked my father if he had any idea what kind of disgusting stuff was on that floor, and his response was that Eric was having fun, so what's the harm? As I said, my father was an American original; you could only roll with the punches.

Eric and Daniel were four years apart in age and had different interests. Eric joined Cub Scouts and eventually Boy Scouts. He enjoyed learning the various items the handbooks taught and especially liked camping, or, more specifically, building campfires. He came home from one camping expedition with the soles of his boots melted. Apparently, his fire got a little out of control! Ellen started a 4-H group called the Ocean Explorers, which had kids of many different ages. Eric and Daniel both participated, and our house was frequently full of kids being kids. One project they embarked on was getting the town to recycle Styrofoam after finding large amounts discarded on one of their beach clean-ups. Ellen worked with Jane, the town clerk, to make sure the kids wrote a legally binding ordinance and collected the correct number of valid signatures. Town meeting at this time was an all-day event where voters would gather at the high school cafeteria to debate and vote on duly presented articles. Usually about a thousand people showed up. While this was a great civics lesson in how democracy functioned, Ellen and I both felt the initiative would not pass. The selectmen and public works department had come out against it because of cost, and the school board opposed it because lunches were served on Styrofoam trays. After a procedural motion passed to let non-voters speak, several of the kids, including Eric, told the assembled voters why this article should pass. Then voters debated back and forth before the moderator put the article to a hand vote. To my amazement, the article passed overwhelmingly! The victory was short-lived, as three years later public works could no longer find a recycler to take the Styrofoam and the ordinance was rescinded. Nonetheless, the results reverberated all the way to the statehouse, as politicians took notice of what the coming generation could accomplish.

Lest people think all we had to worry about was management insanity, we all still had our usual issues of catching fish, fighting the weather, and getting adequate prices for our product. Tom sold the *Sea Fox* in 1989 and went to work running a tuna seiner which had been converted to a dragger named the *White Dove*. A new fishery had finally emerged for spiny dogfish, the schooling shark that plagued my gillnet career. The sharks were being exported to Europe, where they had followings both in Germany and in England, where they were sold as fish and chips. The *White Dove* was a huge boat with very stout equipment that allowed Tom to split the bags of sharks and winch them on the boat. The boat would come in nearly sunk with dogs piled ten feet high after only a couple short tows.

Red and I did very well in the spring and early summer of 1991. We both had new nets built because of mandated mesh-size changes and were catching good amounts of cod. I was exploring hard-bottom fishing while Red continued with the traditional flatfish, soft-bottom fishing. I had a crewman named Charlie who used to say he could clean any number of fish I caught in a given time. Basically, if the catch rate was slow, he would take his time and smoke his cigarettes, but if I put five thousand pounds on deck, he would be like a blur of activity and the fish would be boxed and moved forward before the next haul back. Charlie would be my longest tenured crewman, staying with me for eight years. We had some big days on cod and pollack in early July on the side of Jeffreys. Charlie was in his glory with fish up to his knees. I had found a way through some particularly difficult bottom with my scanner. I now had a plotter, which would draw lines with information it received from the LORAN and save that information so it could be repeated. This allowed me to make turns at precise locations and saved a tremendous amount of mending. My job, other than to keep from ripping the net, was to keep Charlie in lit cigarettes. I would just put them in his mouth, and he would keep on cutting. One day, he needed to do something, and I took my son Eric as crew. Eric never particularly enjoyed dragging, mostly because he tended to get seasick cleaning fish. On this day, there was a big roll from the south and the boat was moving around a lot. I knew Eric was not happy, and I hoped I could get a set of cod and head for home early. We hauled back, and I could see fish floating in the codend, but before we could get it to the stern, we needed to cut loose a bunch of lost gillnet that was wrapped all around the net. Every time I took the boat out of gear to cut the gillnet, the boat would lay in the trough and roll with my net alternately going slack and then tightening like an overstretched rubber band. It was against this backdrop that my net snapped tight while I was sawing on a hunk of the gillnet with a freshly sharpened Dexter gutting knife. As I was applying maximum force, the boat rolled and the knife slipped off the gillnet; I buried the knife deep in my right leg right beside the bone. Instinct is to pull the knife out, but without knowing what it had severed, that instinct could be fatal. Calmly, I told Eric we needed to reel our net, the gillnet and anything else on the reel, put a block on the fish, and get them in the boat before we got the net in the wheel. This we did with the knife still buried in my leg. Next, I told Eric to get the first aid kit as well as paper towels, electrical tape, and duct tape. Then I asked him if he remembered how to use the radio and call for help. Yes, he did. Finally, I asked if he remembered the training in his Boy Scout merit badge about tourniquets and bandaging wounds. This elicited the eyeroll, like, "Of course, do you think I am an idiot?" My real concern was that Eric had always passed out at the sight of blood, but I kept that to myself. I sat on the fish hold cover and told Eric to use the scissors to split my rain pants and blue jeans. This he did, and to my surprise, there was very little blood, but the knife was about an inch and a half into my calf. Eric already had a plan. I would pull out the knife and if blood spurted out, he would apply a tourniquet above the wound. If blood merely ran out, he would apply paper towels held in place by duct tape and electrical tape. I pulled the knife out, and blood oozed out at a good pace, but obviously no arteries had been cut. Eric taped me up and said proudly, "Now we are going home, right?" I responded, "No, we have to untangle the mess on the reel and set the net back out." We would reverse the tow, which took us towards home, and then haul back. He looked at me like I had two heads but did as I requested. He had boxed about fifteen hundred pounds of cod and only had a few miscellaneous fish left when I signaled it was time to haul back. Getting the wire on the winches was not easy, as pulling on it put strain on my leg, which hurt like hell, but we finally got the net in the boat. We had captured a whopping six cod in count. I could not go home fast enough. I helped Eric clean the fish as best I could, but standing up made me dizzy. Finally, I went in to sit with my leg elevated on the dash. We unloaded and I took Eric home. Ellen drove me to the hospital for stitches and a tetanus shot. On the way back from the hospital, I

Left to right, Eric, the engineer, inspecting the winch, with Dan reveling in the fish and smells. Dad trying to unload. Hampton Harbor, 1987.

told Eric how proud I was of him and how well he handled the situation. I was also grateful he had just completed Boy Scout first aid training.

Just as the pain in my leg was subsiding, a new threat appeared. Hurricane Bob looked to be on a course that would take the eye just west of Hampton. This put us in the maximum wind quadrant of the storm. It also meant the wind would come from the southeast. My boat mooring was right near the bridge over the Hampton River and fully exposed to any southeast sea. I made the decision to move the boat to the state fish pier in Portsmouth, as did some other boats from Rye and Hampton. We were all rafted together on both sides of the pier. The only danger was being hit by branches from trees in a neighboring park. Captains all stayed on the boats in case of problems, but the chief problems were boredom and the fact we had no communication with our families. About an hour after the remains of the eye passed, I drove home. My boat was green from being sandblasted with leaves, but otherwise fine. The harbor had several sunken boats and a bunch more that had broken loose and drifted into the marsh. Evacuating the harbor had been a wise move.

The weather, however, was not done with the fishing fleet that year. I was still whiting and herring fishing at the end of October when a new threat began to take shape. A hurricane passed east of us but ran into a huge blocking high over Canada. Meanwhile, a cold front exited over New England and a low formed on the front. From my perspective, this looked an awful lot like the Blizzard of '78, but with energy from a hurricane mixed in. The low would absorb the hurricane but then be pushed backward towards New England by the blocking high. We already had a roll from the hurricane passing by, but as the low pushed back towards New England, the winds rapidly increased from the northeast. Red and I had come in early and loaded what bait we had into my truck. This I parked in front of my house for two fishermen from Portsmouth to collect. This was Halloween night and trick or treaters, including my two sons, were plying the neighborhood. I went out to talk to my two friends as they transferred the bait and was promptly hit in the head with a branch. We discussed the wind, and all agreed sustained winds of seventy knots were possible. They were rightfully concerned about their lobster gear. What no one at the time knew was how long the wind would blow and how big the sea would build. The storm lasted five days and produced seas over thirty feet locally, but a buoy off Nova Scotia reported seas over one hundred feet. The crew of the *Andrea Gale* all lost their lives northeast of Sable Island, Nova Scotia, and the crew of the Coast Guard cutter *Tamaroa* made a heroic rescue of four of five crewmen off a helicopter that had to ditch when it could not refuel while on a rescue mission south of Long Island. Extensive damage occurred from Puerto Rico to Labrador and lobster fishermen off New England lost well north of fifty thousand traps. Sebastian Junger would coin the term "perfect storm" to describe the carnage in his epic book and later movie. However, from my perspective, the damage from this storm, while severe, was not at a level of the Blizzard of '78. In this storm, fishermen retrieved some gear; in the blizzard, they retrieved nothing. All the same, I still have a very hard time reading *The Perfect Storm* and never made it through the movie. The stories hit too close to home.

Fishermen probably should have received a tutorial in civics from the 4-H kids, because their skills at participatory democracy were woefully inadequate. My friend Tom's brother Joe had tried to organize the Gloucester fishermen to participate in a public hearing scheduled for groundfish Amendment 5 in Gloucester High School. The fishermen were supposed to meet in the morning before the public hearing at an Italian social club called the Saint Peter's club.

The government had lost a lawsuit to the Conservation Law Foundation (CLF) and entered a consent decree to produce a management plan that ended overfishing by September of 1992. This amendment was the topic of today's public hearing. Joe and I both felt that by telling the council what parts of the amendment fishermen could accept and offering plausible alternatives to the parts we could not accept, we might salvage something usable out of the process. Our position of compromise was a very difficult sell because many cod had showed up seemingly out of nowhere in the Gulf of Maine, and yellowtail flounder in southern New England seemed to be rebounding as well. This was the backdrop when Eric and I entered the club. Perhaps one hundred men were gathered, liquor was flowing freely, and numerous, vigorous debates were being conducted. My first thought was that it looked like a Tri-Coastal annual meeting. I approached the bar, ordered a Seven and Seven for myself and a Seven Up for Eric. The bartender looked at Eric, who was about eleven, and said if he was old enough to walk through the door, he was old enough to drink and handed him a Seven and Seven. I snatched the drink, went in the bathroom, dumped it down the sink, and refilled his glass with water. This was NOT off to an auspicious start.

Joe laid out the plan we had formulated, which was basically to have the assembled back our positions with a show of hands when called for. Like all council plans, there were many analyzed options presented at the public hearings. Many fishermen did not want anything passed. Joe and I realized retaining the status quo wasn't an option, and so we tried to get the assembled fishermen to back the most palatable option, to which we would then try to offer wording changes to make it more viable. For example, instead of only requiring a hand-written receipt for one pound of fish, require five hundred pounds and a fish house receipt in exchange for being issued a limited-access permit. At least there was no violence at the public hearing, but our voices fell on deaf ears.

The council amendment passed at a later council meeting without having its fatal flaws corrected. Access to the fishery would be limited, but anyone who could produce a receipt for one pound of groundfish could obtain a permit. Two classes of vessels were created: those who got individual days at sea based on actual days at sea and those who got fleet days, which were set at eighty-eight days. This, too, had a fatal flaw, as individual day boats would need a vessel monitoring system (VMS) which would allow the government to accurately count their days at sea but which required more power than boats without generators could create. Thus, the large boats had no reduction in days, while small boats like mine were reduced from about 250 days to 88. After vessels had declared their intent, i.e., individual or fleet, the VMS requirement was dropped for "technical" reasons. There were many other restrictions as well, but as far as I was concerned, this amendment was a complete failure. Amendment 5 went into effect in March of 1994. You could not make a living ground fishing only eighty-eight days per year, so your choice was either to fish those days and spend the rest of your time in other fisheries, such as lobster trap, shrimp, or whiting. Some smaller boats without sufficient capital to invest in these other fisheries went out of business, but most turned their attention to other fisheries and fished groundfish only when necessary.

CHAPTER 20

Managers Wear Multiple Hats

Fisheries are managed by entities in addition to the federal government. The states manage fish and shellfish within state territorial waters and a group of states manages fish and shellfish that are found in multiple states. In the case of northern shrimp, the Atlantic States Marine Fishery Commission (ASMFC) manages shrimp removals, sets the length of the season, and sets gear requirements. Some species like spiny dogfish and herring are jointly managed by the ASMFC and the New England Fishery Management Council (NEFMC), at least in New England. This produces an absurd situation where people who sit on both boards, mostly state directors, bring two hats to the meeting. They will put on their federal hat and say one thing, then change to their state hat and speak against what they just said. You cannot make this stuff up! No wonder fishermen often left meetings scratching their heads wondering what they had just witnessed.

In the case of shrimp, although ASMFC set all the important stuff, the fishery still had to comply with federal groundfish regulations or the fishery could be shut down in federal waters. Groundfish regulations now stated that fisheries using mesh sizes smaller than the groundfish mesh size could catch no more than 5 percent groundfish by weight. So, for example, every thousand pounds of shrimp you caught, if you had fifty-one pounds of combined groundfish you were in violation. Norwegian fishermen had been experimenting with the Nordmore grate system and found it radically reduced catch of fish and shellfish. The system consisted of a grate with a one-inch bar spacing placed at a forty-five-degree angle ahead of the codend. A large, triangular hole was cut in the twine at the top of the net ahead of the grate. A mesh funnel was sowed in the twine ahead of the grate to direct all the catch to the bottom of the grate. The theory was that the shrimp went through the bars and the fish and larger shellfish bounced up the bars and swam out the hole.

The problem was no one in New England had ever seen this setup. Nevertheless, in 1992 you would be required to use one. Now, I am all for bycatch reduction and had even taken several different groups of scientists out on my boat to test their ideas, but all those ideas had been stunning failures. The ideas either did not reduce bycatch or they radically reduced the shrimp catch. What would fishermen do if this device was a radical failure? When you introduce new gear, you need an education campaign. Usually that means building demonstration gear and deploying the gear throughout the range of the fishery to verify it works as advertised, but not this time. You could obtain an information sheet on how to build the device and that was it. I decided to stay ground fishing for the winter, but Red decided to build the setup and give it a try. Red was much more mechanically inclined than I was and enjoyed tinkering with things to make them work. He was also an accomplished welder and decided to build a grate out of stainless steel. His device worked well, and after a winter of tinkering, Red was convinced he had a workable setup. I called my net builder, but he refused to build the device. He did not want legal trouble if something was not installed exactly as the directions stated, so I built one in my basement with my wife holding a protractor to the grate at, hopefully, the right angle while

I hung the contraption from the cellar sewer pipe. We spent days, or more precisely nights, trying to get the thing set at the prescribed angle. Finally, we gave up and hoped for the best. It turned out to be much ado about nothing because all enforcement seemed to care about was the grate and the hole. We would be ready for the next year, but now I needed a separate net for whiting and herring because this device would only work for shrimp.

In the summer of 1993, the whiting dropped off considerably in the last week of July. Whiting is a volume fishery, meaning that

No cooking tonight! David took over the kitchen cabinets and table to build the Nordmore grate for his shrimp net, 1993.

you need many pounds to make money. My crewman Charlie loved tuna fishing and bugged me incessantly to go. I could usually mollify him by pointing out you would need to catch two tuna a week to make what we were making whiting fishing, but that argument did not hold water now. The fishery had changed since the early 1980s when I last fished. Most people had switched to using rods instead of handlines and everyone used monofilament leaders instead of wire. The *Ellen Diane* was not a tuna boat and had numerous impediments to playing a fish which would make landing a fish difficult, if not impossible. However, the thought of going kept Charlie happy, so I put him to work going through my handlines and set off to buy a roll of monofilament and a tool to crimp the leaders.

Our supplies consisted of a casket-sized cooler full of food and ice, six handlines, two snaplines and a dart pole and dart line as well as three boxes of iced bait from the last whiting trip and a bait gillnet to catch fresh bait. Draggers are slow and this would be a five-hour ride. We arrived at sunrise and the area was already well-populated with anchored boats. We did not get a prime spot but got an adequate one. We set one line on the bow, two at the back of the cabin, and two on the stern with a line floated well out behind the boat. This boat was a lot smaller than the party boats and you could not fish as many lines. About two o'clock, I started marking a fish down deep. It seemed to be staying right under the boat. Then I noticed the bow line snapline was stretching back and forth. Finally, it dawned on me, our snapline was really seine twine and the fish could not break it! It had probably been on the hook for five minutes. Charlie went up and cut the snapline and the fish took off. We played the fish for about twenty minutes before it spit the hook out. It had probably torn a hole in its jaw trying to break the snapline. We were discouraged to lose the fish but excited to have some action. We rerigged and found some appropriate snapline. Around five o'clock there was a tell-tale snap and one of the cabin lines began streaking out. This went more according to plan, and we soon had the fish darted and tail wrapped. The fish did not put up as much fight as I remembered and after loading the fish in the boat, I figured out why. The fish looked tiny to me. In my previous tuna life, the smallest fish I caught dressed at 525 pounds. This fish would dress at 400 pounds. I called my wife on the citizens band and had her tell the coop we would be in Hampton about midnight. I also asked her to bring a thermos of coffee and a pizza. The fish was out of the boat, and we were headed back around 1:00 a.m.

I had been up over twenty-four hours and my body was complaining. The coffee kept me awake but gave me a headache. I told Charlie I would stand watch until we were east of the Isles of Shoals and then he could take a couple hours. I explained the radar and said if any target showed up within the two inner-most rings to wake me up. I slept for about half an hour before starting to toss and turn. I have never been able to sleep on a boat, mainly because I never trust the person I leave on watch. Just as I was about to get up, Charlie appeared and said there is something out here. I darted out of the focsle to be greeted by a red light at close range. Charlie was about to drive between a tug and barge. I stopped the boat and thought about berating him for not following my instructions but decided it would not change anything. I drove the rest of the way. When we arrived on the grounds, I had a hard time finding my anchor. At first, I thought it had been cut off or someone had hauled it and moved it, but finally figured out there was a skiff anchored nearly on top of the ball. I picked up the ball and fell back on the anchor and went to sleep.

When I woke up the skiff was gone. I guess looking up at my trawl door hanging above his head caused him to relocate. This day we did not have to wait long for action. We hooked up at 9:00 a.m. and had to let the fish drag us out of the fleet. After untangling the line from several anchor balls, we fought the fish to the surface in the deep water. Dragging it back to our anchor, I was amused to listen to the radio. I had equipment that could listen to every device someone could talk on, and there was plenty of talk. I had two fish in two days and most of these people probably did not catch one fish per year. I had not fished tuna in ten years, and there had been a generational shift in the cast of characters. People were asking who was the old guy with the dragger? That made me smile because a few years before fishery management people called me sonny. Finally, someone piped up and pointed out I had caught more tuna before they were born than they were likely to catch in their life. That made me smile, too, because it was probably true, but the fact was, I had never fished this ground before and had picked my spot based on availability.

This time, I had Charlie haul the anchor. The weather service had put up a tropical storm warning and now was the time to beat a retreat. This fish would weigh 350 pounds dressed. We had the next day off while the tropical storm went by but left the following night at midnight. I had seen weather events radically alter the fishing before and I told Charlie we would fish half a day and if there was no action we would go home and return to whiting fishing. All the herring and small pollack that had attracted the tuna were gone and nothing was happening. No one had a bite by noon time when I started the engine. Charlie thought I was nuts, but I hauled the anchor and departed. There would only be two more tuna caught in that area for the rest of the year. Our fish were high-quality, and we received ten dollars per pound for each one. Charlie bugged me the rest of the year to go again, but I believe in leaving the table when you are winning.

My friend Marty, who owned the moving company I had worked for, called one night to ask if he could bring his son out to see how dragging worked. His son was about six and clearly had a huge interest in fishing. He thoroughly enjoyed every aspect of the day, especially the big catches of whiting. That little boy grew up and will now be known to readers as Tyler Mclaughlin, captain of the *Pinwheel* on the TV show *Wicked Tuna*.

Meanwhile, my two boys were growing up. Eric had taken a very serious interest in flying at the local airport. When he was fourteen, he obtained a job as crew on a party boat in Rye. He spent all his money on flying lessons. Daniel, on the other hand, had developed a keen interest in sports, specifically baseball, basketball, and soccer. He had sufficient talent to be placed on traveling teams, which as the name implies means you travel to play other teams outside your local town. The mom cab seemed to be in perpetual motion.

In addition to all this, Ellen had started her own business of taking sea animals to schools to teach children an introduction to marine science. This produced some odd and unexpected stories as Ellen kept live specimens in a lobster tank in the garage and dead stuff in the family freezer. Much of our frozen food had a special scent, best described as eau de sea star. Ellen was fearless and always on the lookout for new specimens. One day, she decided to remove an eye from a tuna head at the coop. This turned out to be harder than she thought, and a crowd soon gathered to watch this strange woman excavate the head. The Japanese graders seemed particularly interested and spoke excitedly in Japanese to each other. Ellen could not figure out whether their excitement was caused by the oddity of what they were seeing or if this woman might have a recipe for eyeballs. After about twenty minutes, the operation was a success and the eyeball dropped into a jar of preservative. Ellen brought home her trophy and for some unknown reason placed the jar on top of the downstairs toilet. Eric came home from school, walked into the bathroom, and promptly let out a blood-curdling scream while fleeing the bathroom. "There's an eyeball in there, you cannot use the bathroom with an eyeball staring at you!" So, Ellen put the jar in the refrigerator. Daniel walked in, opened the refrigerator, and filed a similar complaint. "How can I eat with an eyeball looking at me?" The jar ultimately ended up on a shelf in the garage. Our house was like a museum; Boy Scouts, 4-H, and neighborhood kids all wanted to see what Mrs. Goethel had been collecting.

Meanwhile, with the ink barely dry on Amendment 5, in 1994 the New England Fishery Management Council (often just called the council) began work on another major change to fishery management. The goal of Amendment 7 was to cut fishing mortality by 80 percent in two years. This meant, for example, if the total catch of a species was one thousand tons of both landings and discards, at the completion of the amendment both landings and discards must be 80 percent less, or two hundred tons. The major components were complete closures of vast pieces of Georges Bank and southern New England waters, accelerated reductions in the number of days vessels could fish, closure of all waters to small-mesh fishing that could not prove those fisheries caught less than 5 percent by weight of groundfish, and splitting the commercial and recreational fisheries. NMFS also had a change in leadership in New England. Dr. Andrew Rosenberg assumed the post. He believed strongly in closed areas as a fishery management tool and had been given the position to get tough with fishermen in New England. New England had a bad reputation both with fishcrats in Washington and with some of President Clinton's senior environmental advisors in the secretive executive branch, Council for Environmental Quality. New England fishermen, and to some extent senior politicians, had what from Washington's perspective was a bad habit of not blindly drinking whatever Kool-Aid was dispensed by the bureaucracy. Public hearings were duly held, but questions raised, such as what happens when all the large boats shut out of Georges Bank flood the Gulf of Maine and how do you prove your bycatch is less than 5 percent, went unanswered. The amendment went into effect in the late spring of 1996. This was the period of amendment by lawsuit. Nothing was given time to see if it solved problems or created new problems, just get something in place that works on paper by a certain time.

This amendment presented an existential threat to small boats in the western Gulf of Maine. The gillnetters like Erik would have their gear mauled by the big boats which towed large rockhoppers that could go places the rest of us never ventured. The shrimp fishery had been declared bycatch acceptable so that was good news, but the whiting fishery was arbitrarily shut down. Draggers could not survive on shrimp and what meager days they were allocated for ground-fish. My friend Tom had bought a new larger vessel named *Midnight Sun* which he planned to use for whiting, but that fishery was now closed in the Gulf of Maine. Amazingly, southern New England, which has extensive small-mesh fisheries, was declared bycatch compatible after a Rhode Island Council member had threatened to vote against Amendment 7. As always, the politically powerful marched on and the little got steamrolled. I held meetings with Tom's brother, Joe, who had emerged as the de facto leader of the Gloucester whiting boats. We had both seen this threat coming the previous year and tried to enlist state biologists to go out with us to document bycatch. We also took people from the newly formed observer program on a volunteer basis and convinced some of the older men, who generally distrusted anything run by the government, to participate for their own self-preservation.

Through their fish and game departments, the states of Massachusetts and New Hampshire both tried to intervene with Dr. Rosenberg by using the col-lected observer information, but they were rebuffed. So, we did what citizens do, contacted our congressmen and pled our case. My congressman, Bill Zeliff, wrote a letter to NMFS and received a one-page response that basically said no, strong letter to follow. My junior senator, Judd Gregg, sent a letter and received a similar response. However, my senior senator, Bob Smith, who was strongly anti-big government and chair of the senate committee that oversaw the NMFS budget, had his staff write a bluntly worded letter suggesting Dr. Rosenberg find time to meet with his fishermen or he would hold hearings to zero out the NMFS budget. The response came back, how may I help you, where, and what time? The meeting was held, with state biologists, Senator Smith's staff, Dr. Rosenberg, Joe, and me. We brought charts and laid out two areas where we were certain we could meet the 5 percent threshold during certain times of the year based on the biological data collected. We had to forfeit some areas and time because we had collected no data in those areas, but at least we could survive. Joe and I emerged as formidable adversaries and as a new breed of fishermen. We argued with scientific facts and lined up political clout to force the bureaucracy to listen. NMFS also learned that just saying no was not a viable strategy. This meeting also started a tradition where Ellen and I would brief our congressional delegation on fishery issues. We did not lobby, but rather presented both sides of the issue, and only if asked our opinion would we provide it. We offered this service to both sides of the political aisle and never framed the discussion in political terms. We never violated anyone's trust, and no one ever violated ours. This tradition continues to this day.

September 5, 1996, dawned foggy and warm. As I left the harbor, I was immediately struck by the amount of radio traffic on many different channels. Tuna fishing had moved to a "days off" management model, and when it was open there were numerous boats on the water. Much of the traffic was captains trying to contact vessels on crossing paths, but one call caught my attention. There had been a collision on Jeffreys Ledge between a tug towing a barge and a boat. Details were sketchy and monitoring the traffic proved difficult because so many people were trying to contact vessels on the radio. The Coast Guard was involved and was trying to gather information and dispatch a rescue boat. After listening for a while, I determined the boat involved was the *Heather Lynne II*, which was a gillnetter from Newburyport owned by a colleague from Tri-Coastal. The boat had run between a tug and a barge and been flipped over by the towing cable. The *Heather Lynne* was upside down, still floating with men trapped inside. At least one person was tapping from inside the upside-down hull. The Coast Guard was waiting for a dive team from Beverly to arrive to transport them to the scene. I had an analog cell phone and called my manager, Bob, who in earlier days had been a crewman on the boat. I knew time was of the essence and figured Bob might know some divers in Newburyport.

The first Coast Guard boat to arrive forbid anyone from attempting rescue until the dive team arrived. The boat had been carrying gillnets, which were floating all over the surface, and the Coast Guard was afraid of people becoming entangled. Tempers on the scene flared, radio communications with land were broken at best, and several people basically were forced to stand by and do nothing. By the time the Beverly dive team arrived, they were too late. The men had all drowned. The fishing community was stunned and very angry. Several people had offered to free dive and begin clearing away the nets. Perhaps it would not have mattered, but the idea of sitting around doing nothing was unacceptable. Kate Yeomans, a family friend, would chronicle this tragic day in the gripping book: *Dead Men Tapping*. Years later, this tragedy still haunts all involved. Several people and groups, including the Coast Guard, would be held accountable and practices and procedures were changed, but none of that brings back the dead or provides solace to the families. This day was a tragic reminder, that even on calm days, fishing can be a deadly profession.

CHAPTER 21

Frameworks and Amendments, Oh My!

Regulations continued at a frenzied pace. The council met every other month and needed a faster way to pass regulations. The council started to employ its framework process, repeatedly skipping amendments which took much longer and required more public input. A framework could be initiated at a council meeting, and if the staff could complete the analysis by the next council meeting, a final vote was taken roughly two months later. Because council meetings have limited time, public comment was severely restricted, often limited to several people for or against the framework. In rapid succession, a new net was required for the whiting fishery, logbooks were required, trip limits for cod were instituted, and a huge piece of the western Gulf of Maine was closed to commercial fishing but left open for recreational, charter, and party boats. Every one of these items came with huge, negative, unintended consequences which were never explored or discussed because of time constraints. The system was broken, and no one wanted to take the time to have a rational discussion about how to fix it.

Throughout these dark days, people continued to catch fish. In April of 1997, Red had two tows which netted what I believe was the largest amount ever of dragger-caught cod in Hampton Harbor. He called me with detailed instructions of where he towed, but I never found the fish. Red landed just a shade under ten thousand pounds while I had about three hundred pounds. That is why it's called fishing! Erik had located a school of cod in May that was producing exceptional catches. I stopped to talk with him one afternoon and asked if he could let me tow through the area once while he had the nets in the boat. I lined up the tow and went down between the two humps he was fishing on, hauled back, and had over three thousand pounds of cod. This was the type of cooperation that kept us all in business. We were a community on the water and looked after each other.

By the start of the whiting fishery in mid-July, Tom and I were trying to figure out the complexities of the new raised-footrope trawl. I knew the researcher who had done the experimental work from attending several responsible fishing forums, one in Scotland and the other in Saint John, Newfoundland. I also knew that the area where the work had been done off Provincetown used very different nets than we did. We used nets with adjustable disc sweeps with very thin-diameter ground cables to keep from catching bottom-tending species. This new net required a high number of floats and an extra-long, thin chain sweep that allowed the net to travel off the bottom. This was attached to large-diameter wire cables. When we rigged our nets and attached them to our thin ground cables, we caught virtually nothing. If we cut off enough floats to catch anything, we were in violation of the law. We felt the whiting and mud hake were going under the bottom of the net through the chains the same way they went through rockhopper sweeps. Maybe the fish in Ipswich Bay had a different behavioral response than those in Massachusetts Bay. I talked to the researcher every day and came back with new things to try and to measure. Tom had some of the heavy cables the researchers used left from his days on the *Sea Fox* when Eastern rigs used cable and wound the ground gear on the main winch. He put them on and started catching fish. It turns out the heavy cable keeps the ends of the net near the bottom. The thin cable we were using was too light, and the net lifted the cables off the bottom and traveled too high in the water column. Once we procured the necessary cables, the catch rose dramatically. This again was a case of not showing people how to use the new gear or fully understanding how each component of the gear contributed to the catch.

My kids were both teenagers now, with all the joy and angst that period in life provides. Eric flew a plane solo on his sixteenth birthday. This is a huge achievement for a young pilot, as it means the instructor believes he can receive his license. I am still surprised you can fly a plane before you can drive a car. Daniel was beginning his teenage growth spurt, and one could see he was going to be very tall. His interest in baseball waned but was replaced with intense efforts in basketball and soccer. Daniel had a very keen sense of fair play and expected referees to keep the playing field level. At this age, however, the referees are often kids themselves and do not always see everything. We had to warn his coaches to take him off the field if more than two egregious fouls went uncalled. He never got mad, but when he decided to get even someone could get hurt.

Summer of 1997 changed to fall and fall to winter. We shifted fisheries accordingly, going from whiting to herring, back to groundfish, and finally to shrimp. Each came with their own challenges and rewards. The council process continued as well, but all it ever provided was challenges. Another year, another bad cod assessment. It just boggled the mind. How could this government research boat, which conducted the trawl surveys that fed into stock assessments, never catch any number of cod, when we could not get away from them? Council meetings were becoming downright dangerous. Federal law enforcement and local police now routinely attended meetings, making sure everyone could clearly see their holstered firearms. This is what we had come to; someone possibly getting shot over a fish.

Against this background, I received a call from a writer. His name was David Dobbs, and he was researching a book on the differences in perception of cod stocks seen by fishermen and scientists. I had developed good relations with press people because I felt the story of how we feed America had been lost in the anti-fishing hysteria created by the ENGOs. I found most journalists, as well as television reporters, were honestly trying to show America this story. However, I had been recently burned by a television crew from a small station in central Massachusetts who filmed the regulatory discard of cod we were forced into daily and spun the story as a wanton attack on a nearly extinct fish. I was furious, but under the First Amendment they could say whatever they wanted. I quizzed David intensely, and he seemed to genuinely be interested in the entire story, so I agreed to take him out. That first day turned into a several-year exchange of information. He fished with me numerous times, even though he was prone to seasickness. I had to admire someone who would spend a day in misery to get the story straight.

In 1996, the Magnusson Act had been reauthorized by Congress and several changes enacted. The new law was much more prescriptive and left NEFMC members with little room to maneuver if they could not prove they met scientists' objectives. The 1996 Magnuson reauthorization had made fishery management all about meeting numerical goals. The management council had to prove, at least on paper, that its plan had a high-percentage chance of meeting its stated goals. Fishermen, for the most part, had little understanding of the law and often felt council members had it out for them. My friend Erik had been appointed to the council as the obligatory representative from New Hampshire. Fishermen felt this was a hopeful sign, as now someone was there to explain what fishermen were seeing on the water. Erik soon learned that the new law was prescriptive, and he had to vote for an alternative that satisfied the law, even if it made no sense. Basically, the council's job had become deciding whether fishermen should be shot, hanged, or poisoned.

For my son Eric, 1997 was a momentous year. After earning his pilot's license, he would rent planes from the airfield as frequently as his budget would allow. When you asked him where he flew, his response was always, up, around, and about. Eric would also receive his Eagle Scout award from the Boy Scouts in the fall. This was a huge milestone for Eric personally, but also for the town and troop as few young men achieve Scouting's highest rank. Eric built a walking trail in the woods adjacent to the elementary school behind our house in Hampton with help from sailors on the USS *Hampton*. The trail is still in use today. Eric would graduate high school in the spring and prepared to attend Boston University's College of Engineering.

By the summer of 1998, fishermen braced for the annual report of the spring research cruise of the government's research boat *Albatross*. Once again, they found few cod and virtually no recruitment, which is a scientific term for the birth of baby fish. More cuts were coming. The meeting to finalize those cuts would be held in Portsmouth, NH, in late January at the Sheraton Hotel and would become known as the midnight massacre. Council meetings usually last three days and must be posted in the Federal Register well in advance. Groundfish is usually allotted one full day, as it is the most contentious managed species. Meetings usually start at 9:00 a.m. and are usually concluded by 5:00 to 5:30 that afternoon. Having attended and participated in council meetings for years, I always found the quality of decision making fell off markedly as people became exhausted by the process late in the day. I met David Dobbs at the meeting, and we exchanged pleasantries, and he asked me to handicap the alternatives. I responded that none seemed viable to me, but my assumption was the inshore boats would likely get thrown under the bus because of the council make up (New Hampshire has always been represented by small-boat fishermen, but not so other states). Massachusetts and Maine especially seemed to be influenced by the large boat ports of New Bedford, Boston, and Portland. I asked how his writing was coming, and he said he was progressing well and that the outcome of this meeting would probably signal the conclusion of his book. I wish to state here that I went back and reread the book to see if my millennial self had had the same views I hold today. I was surprised to find that, yes, my views are essentially the same. I still believe that scientists and fishermen are two sides of the same coin and if someone had had the foresight to force the two groups to work together, an awful lot of fishery mismanagement could have been avoided. I also think that many of the problems stemmed from bureaucratic inertia, combined with plain stubbornness and unwillingness to change on the part of a minority of NMFS scientists, was a large part of the failure of the process. I was not alone in that view; it was shared by several scientists portrayed in Dobbs' story.

As the meeting began, snow was falling at a good clip. Hell was truly freezing over. By the middle of the afternoon, the council had debated all three of its staff-analyzed options and voted all three down. Absent a motion to reconsider from the prevailing side of one of the motions, the council was headed for uncharted and dangerous territory, crafting an unanalyzed motion on the fly. After a brief recess, hybrid motions were put forward. Debate raged, items were placed on the board later to be withdrawn, tempers began to fray, and the clock ticked on. Darkness fell and still nothing even remotely approaching consensus appeared. The collective IQ of the council fell at an alarming rate; this was going to be ugly. The audience was also growing restless. With no motion to speak to, they were reduced to potted plants, just sitting there waiting for something to happen. Finally, a fishermen stood up to make a point, recognized or not, and walked rapidly down the aisle, past the public microphone into the horseshoe shaped tables that held the council. He began screaming at everyone as his frustration had boiled over. Enforcement officials and Portsmouth police approached the man and told him he had to leave. They were very polite, almost apologetic, but emphatic that he had to go. For a moment time stood still, everyone waited to see if the man would comply or continue moving toward the council chairman, in which case, we would all be witness to a shooting. The man just stood there considering his options. Finally, he stood down and left the room. The council looked deeply shaken and decided to take a five-minute break. This was the first thing they had agreed on all day. I always believed that the meeting should have ended at this point and have been resumed the next morning after a good night's sleep, but the chair insisted they soldier on. By 9:00 p.m., they still had no agreement, and the executive director was informed by hotel management that if people wished to eat, they would have to do so now. This created the final fiasco, as the only place to eat was the lounge. An hour dinner break was declared and the whole meeting poured into the lounge. What could possibly go wrong? An hour later, a considerably smaller group returned to the meeting room. Many of the audience members, having looked outside, had decided to call it a night and go home. The council put forward a motion for massive, multi-month closures with a two-hundred-pound, cod-trip limit which could be altered down to any number, including zero, by the regional administrator of the National Marine Fishery Service. The inshore boats like me were dead, both literally and figuratively. Our boats were not big enough to travel safely around the closures and two hundred pounds of cod would barely pay for the food and ice. Very limited debate ensued and an even shorter public comment period. Council members, by this time, would have held ritualistic sacrifices of their own mothers if it would have ended the meeting. A little after 1:00 a.m., the council passed the motion as written. Only about thirty of us stayed until the end to witness the midnight massacre.

CHAPTER 22

Midnight Massacre, the Aftermath

The effects of the midnight massacre began almost immediately. My crewman Charlie told me he would stay through the end of shrimp season but had already begun looking for work on land. Small boats that held lobster-trap licenses would return to that fishery and trip boats would return to Georges Bank, which had been conveniently expanded twenty miles to the north into the Gulf of Maine by a trip boat council member from Maine. Georges Bank had also conveniently had cod-trip limits raised. I would fish alone and take the captain and crew share. This was the only way we could possibly survive. The plan was to go north in April to an open area that I knew would not have many fish and go south forty miles in May to an open area in that month. These would be twenty-hour trips with just me on the boat. I had refused to take my father on day-trips the year before as his physical strength and cognition were both declining. He was simply too old to be safely wandering around the boat. My friend Tom provided me with a list of wrecks and hangs in the open area east of Boston which helped tremendously. Losing a net and doors right now would have been catastrophic. Ellen told me to begin documenting the unnecessary slaughter of cod above the trip limit and write letters of complaint. My wife was extremely displeased to have me fishing alone and was not shy about telling people so. She had always been a force of nature when someone threatened the family unit, and she felt like we were having a gun held to our head. Eric was away at college having scored a 50 percent merit scholarship to attend BU if he maintained an A-average, but Daniel was taking this all in. Daniel was a quiet, shy, young man who spoke very little to anyone, but his hearing was top notch and he missed none of our discussions, even when we thought he was asleep.

April was the disaster I expected. The open area had very few fish and my 50 percent share was barely enough to buy the groceries. My plan for May was to fish out of Gloucester, unload there three or four days in a row, and sleep on the boat. That plan was abandoned the first night, as Gloucester Harbor is about

On a cold winter's day, Yankee Fishermen's Cooperative, Seabrook Harbor, NH, 2000.

as quiet as a rock concert. My new plan was to leave Hampton at two in the morning, arriving to fish around six, make tows until 3:00 p.m. and steam back to Seabrook to unload. I was now selling fish to the Yankee Cooperative, which was built on the former reactor barge unloading facility on the Seabrook side of Hampton Harbor. In an outbreak of common sense, the state had forgone the requirement for the power plant to remove the barge facility if they built the fishermen an unloading facility complete with coolers, icehouse, and fuel systems. The power plant saved about half a million dollars, the fishermen got a state-of-the-art facility at no cost, and my former manager at Tri-Coastal had been hired as manager. A win all the way around. The coop would stay open to unload the fish and place them in the cooler until morning.

Fishing in May was grueling, as the four-hour ride in the dark with nothing but a thermos of coffee for company took its toll. The weather forecasts were about as accurate as the fish forecasts but judging the sea state in the dark was nearly impossible. More than once, I got thrashed in weather conditions that were not suitable to my size of boat. The area I fished was my old tuna stomping grounds, the northwest corner of Middle Bank. The area still had prodigious quantities of life, especially whales. One day I found myself surrounded by humpbacks. I find it nerve-racking to have six whales breaching and bubble feeding within thirty feet of the boat while towing. You know your heart still works well when you open the door on the side of the pilothouse to be regarded by a huge eye swimming along beside you less than three feet away. A lot of days, the whales were the only company I had. The northeast wind was the bane of my existence down here as the course home was thirteen miles to the north-northeast, then around Cape Ann, and twenty-one miles to the north-northwest back to Seabrook. Several times per month, I would get the living daylights beat out of me trying to get home. I was lucky to have a boat that handled this sea well. Most days, I threw over four to five hundred pounds of cod for the day, but one day the machine showed lots of sand eels on the bottom. What I could not discern was that the cod were eating them. I did not tow long because dogfish had been reported in the area the day before and a big bag of them could be catastrophic for one person to handle. On haul back, the end of the net blew out of the water before the trawl doors were even to the surface. Well, at least they weren't dogs, which do not float, but the alternative was even more disheartening. Approximately ten thousand pounds of cod were in the net. I was fishing in shallow water, so barotrauma was minimal. I pulled the end of the net around, pulled the tripper and let the whole set swim away. I gaffed up two hundred pounds of fish that were floating dead on the surface. Most of the fish lived, but that was not the point. This management plan was a crime against nature. Every fisherman I knew espoused a philosophy of you kill it, you keep it. This was like a deer hunter who wounded a deer and was too lazy to track the deer down and finish it off. The government philosophy was you kill it, you dump it back dead to rebuild the resource. I did not have a camera, but I did have a four-hour ride home to compose the letter I would write in my head.

The next morning, the letter poured out onto paper at my kitchen table. I was surprised the table did not spontaneously combust. At my best, I found the language of diplomacy to be a waste of words, and at my worst I was described as a loose cannon. I was as mad as I have ever been, but not stupid mad. Tell people what you really thought, and no one would read it except probably law enforcement who would charge you with threatening government officials. Make the letter too saccharin and no one would care. I struck a balance and had Ellen read it. She could edit me and take the loose out of my cannon. I mailed the letter to every NMFS official, council member, the governor, and my congressional delegation, as well as the press. I never received an official response but heard through my intelligence network that I (or someone who had received the letter) had severely rattled NMFS's cage. Senator Judd Gregg would soon appropriate money for cooperative research, scientists and fishermen working together to solve scientific questions, but in the meantime, NMFS lowered the trip limit for cod to thirty pounds on June 1 because 50 percent of the cod quota had been landed in two months, even with the two-hundred-pound trip limit, and most of the ocean was closed. Go figure.

David Dobbs came out fishing with me in early June, both to do some fact checking on his writing and to see the results of the midnight massacre in action. I was doing reasonably well on flounder, but still catching three to four hundred pounds of cod per tow. I gave Dave a tutorial on how to release cod to give them the highest chance of survival. Basically, I kept the hose running in the bin to keep the fish's gills wet, grabbed them just forward of the tail with a wet rubber glove with one hand, and supported their gut cavity with the other, finally dropping them headfirst into the water. Speed was of the essence, so this was quick and dirty. Grab the ones that are obviously alive first and work your way through the pile. Many of the ones that appear dead are only stunned and will come back to life. Finally, pick up however many dead ones you think make thirty pounds and throw the rest over. I saved several fish for the end that were in an obvious stage of decomposition and told Dave this was our government's legacy: an ocean full of dead fish in the name of conservation, but, hey the ENGOs were happy.

Every fisherman I knew was demoralized and mad. Not only were we all working alone, one person doing the jobs of two to three people, but we were downwardly mobile. Captains could make up some of their lost income by claiming the crew share, but the total pie was much smaller, and the boat share was not covering expenses. Ellen and I had both come from modest means and had saved instead of splurging when times were good, but many had not, and thus had little or no reserves. Dark tales of social ills circulated on harbor town waterfronts; it seemed everyone knew of some family succumbing to one or another. My son Daniel was most affected in our family. He overheard the discussions about finances and saw the middle-class perks like family vacations, meals out at restaurants, and even favorite foods chopped from the budget. Daniel was one of the top students in his class and took all advanced classes. He had enormous amounts of school homework as well as a three-season sports load, so I was not surprised to sometimes see him at the kitchen table writing a paper when I left for work. However, one morning near the end of June, he was sitting at the kitchen table eating breakfast, clearly not dressed for school. I asked him what was up, and he responded that he had been watching me, as he put it, "kill yourself throwing over cod" and he was going with me to help. I was both stunned and incredibly sad. Fourteen-year-olds should not be up at 4:00 a.m. to keep the family business afloat. I figured this would last a couple days, but instead he fished with me every day he was not in school for the next fourteen years. Daniel had this job figured out in a week, and by that time, it seemed like he had been onboard forever. Numerous parents in the coming years would ask me to take their child as crew, but this is considered dangerous work and non-family members could not sign on until they were eighteen. Daniel put all his money in the bank to help pay for college. When we switched to whiting fishing in July, we worked together; the two of us would catch, load, pack and deliver up to seven thousand pounds of whiting plus bait and be in by 2:00 p.m. Having Daniel on board were the best days of my life.

Can I interest you in a whiting? Dan Goethel, Hampton Harbor, NH, 2004.

Haddock lost in a school of whiting carefully returned to the ocean by deckhand Dan Goethel, 2003.

Daniel returned to school in September of 1999. Luckily, I was able to locate a man who had lost his job on a gillnetter from Portsmouth who went with me through the winter. About the same time, the council received a grant to form a research steering committee made up of fishermen, council members, and government and academic scientists. Congress, strongly influenced by Senator Gregg, had appropriated money for cooperative research, i.e., scientists and fishermen working together to reach consensus on research questions. As he put it to me in discussions on the subject, the two groups had mutually exclusive views on the stock status of cod.

Both groups had turned to Congress to referee the dispute. Congress had no knowledge on the subject, so it would pay for us all to come up with an answer. This was a lofty goal and probably undoable in the timeline provided, but it was a healthy start on a perennial problem. The first meeting went about as expected, with lots of venting from all camps, but soon a strange dynamic emerged. The scientists argued more with each other than with fishermen and the same occurred amongst fishermen. Also, state bureaucrats who thought this was going to be a money grab soon learned that projects would compete on merit and bang for the buck, and that fishermen could be very demanding financial stewards.

Northern shrimp, managed by the ASMFC, had a separate research cruise and assessment run on the vessel *Gloria Michelle* during the summer. This vessel was used to survey the northern shrimp population and only towed in the Gulf of Maine. The other research vessel mentioned earlier, the *Albatross IV*, made two surveys, one in the spring and one in the fall towing computer-generated stations in different depths from Cape Hatteras, North Carolina, to Canadian waters in the Bay of Fundy. Unlike the cod assessment, the shrimp results tended to track well with what fishermen saw on the water. So, the news that the assessment was poor and the season would be short came as the final blow to many dayboat fishermen. Many decided to haul out their boats for the winter and take whatever jobs were available on land. I decided to stay fishing, as I had a crew and there would not be much competition fishing for shrimp. For once the survey was wrong. The price climbed to a dollar per pound and the catch rate was up to five thousand pounds per day. The shrimp were close to shore where you could fish, albeit very uncomfortably, in the winter southwest and northwest gales. I pushed the weather hard and at the end of the season had enough money to skip fishing in April and do some much-needed maintenance. Throughout my career, I had always found that when life seemed the darkest, nature always provided. I had promised my wife when we married, she would never be rich, but she would also never starve. We had now seen both ends of the equation together. Another old saying is God watches out for drunks, fools, and fishermen. I am not sure what order they fall in, but maybe we were some versions of all three.

David Dobbs' book, *The Great Gulf*, would be published in early 2000. Like all authors, he embarked on a book tour, which often meant discussion groups with fishermen and scientists. A version of this had been on Boston Channel 4 and it was awful. A prominent NOAA scientist had debated one of the more colorful fishermen from Gloucester. It came off more like *Fight Club* than a reasoned discussion. I would later learn from David's book that the scientist had been coached in debate strategy. Winning was more important than constructive discussion to NMFS leadership. I was invited to participate in one of these discussions at the New England Aquarium in late winter. The event would be held on the aquarium's marine mammal show barge, the *Discovery*, which is moored adjacent to the main building. The participants would be David, a professor from the marine biology program at BU, and me. I was all set to accept, but Ellen was worried. She had seen the debacle on television and was afraid I could easily be baited into a verbal tirade. David was worried as well and had secured a guarantee of a neutral moderator to ask the questions. We would all have an early dinner with senior aquarium officials and then proceed to the forum.

The day of the event, just as we were leaving the house, David called. NMFS, at the very last minute, had told the aquarium they were sending the head of the Northeast Fishery Science Center (NEFSC) to the discussion. The aquarium had accepted without checking with anyone else. I could back out if I felt sandbagged. Ellen and I held a quick discussion; my feeling was the fishermen would get worse press if I was a no-show than would occur if I got ground up in the minutiae of assessment science. Besides, I am an optimist, someone might even agree with me! The dinner was humorous. My old boss, Al, who was now director of the research department and gregarious as ever, was more than happy to tell everyone about our adventures in the bowels of Boston twenty-five years earlier. Most of the rest of the aquarium leadership had no idea who Ellen or I were, but since we were obviously alumni, they offered moral support. I had a chance to introduce myself to both scientists briefly. Steve from the science center and I had interacted before about problems with the cod assessment. At that time, he had agreed I had found valid issues in the nearly six-hundred-page report. Les, the PHD from BU, had done his studies on fish in Africa, and he had no firsthand knowledge of cod. He clearly thought the ENGOs were right but did not appear confrontational. Besides, I was a BU alumnus as well, and my oldest son was currently helping to pay his salary. About this time, wine bottles were making numerous rounds. Mental note to self, leave the alcohol for the after party; I will have another glass of water, please.

As we crossed Atlantic Avenue back towards the aquarium, I saw hundreds of people streaming onto the *Discovery*. I would guess the barge seated four hundred people and appeared nearly full. In the front row were the NMFS Regional Administrator and all her senior staff, but seated behind them were two of my former crew from the party-boat era and their wives. Trying to scan the audience through the glare of the spotlights, I guessed I at least recognized half the attendees. What happened next probably surprised everybody. David described his research into writing the story and how he still could not reconcile the two very different pictures of the Gulf of Maine. Steve went next and described logbooks he was currently studying from haddock fishermen in the 1930s and the insights their writing provided. He described Henry Bigelow's work at sea and said he felt the science center needed to get back to its roots of working cooperatively with fishermen. Encouraged by what I heard, I laid out briefly why we needed more research into understanding cod movements and stock delineation, as those had major effects on assessment outcomes. I then stated that could best be accomplished by cooperative research. Les ended up the odd man out, using his time to push for making sure overfishing never occurred no matter what the cost to fishing communities. We all answered questions for over two hours on numerous subjects. Those who came to see a car wreck or fight club left sorely disappointed. For once, I was part of a thoughtful and intelligent conversation about a very serious issue. By exposing the deep divisions and sometimes boorish behavior on both sides of the issue, David's book may have served as the catalyst to pull everyone back from the edge of the abyss. The afterparty would be fun, after all.

CHAPTER 23

Never Order Pancakes

One cold, raw, June day, Daniel and I stopped for breakfast at the luncheonette at the top of the pier. I did not like the feel of the wind and told Dan we would wait for daylight. The luncheonette was run by a woman named Ute, who had immigrated from postwar Germany and was known for her salty language and good food in large portions. Her restaurant mostly sold breakfasts to the people taking trips on the party boats, but she had become well-known on Hampton Beach and now catered to a year-round clientele. Her walls were decorated with colorful, some would say obscene, bumper stickers, skewering mostly politicians but also famous people known mostly for being famous. One of her favorites was: "Why do they call this tourist season if you can't shoot them?" She sold tee shirts with this stenciled on the back, and I happened to be wearing one that day.

Ute brooked no dissent from anyone and could easily pin back a recalcitrant set of ears with language that would make a sailor blush. The language was probably enhanced by the bottle of vodka under the counter from which she periodically fortified her coffee some mornings. The fact was, underneath the gruffness lay a beach treasure who quietly fed people down on their luck and collected presents for people in the nursing home in Hampton every holiday season. Regular customers knew, but she neither sought out nor wished for publicity. One thing that set her off was people inquiring about the food portions. An inside joke was to steer those people to ordering pancakes. This morning, there was a guy in the restaurant either still drunk from the night before or getting started for the morning. One of the fishermen, nursing a coffee, quietly suggested he order the pancakes. Now Ute's regular meals would overstuff the most prodigious appetite, but she served pancakes on steel-bar trays that were somewhere around two feet in diameter. There were two pancakes, about three-quarters of an inch thick covering the tray. She would warn them, "If you do not finish, you are mine for the day." No one knew exactly what that meant, but we had seen tourists drop a twenty on the bar and run out the door rather than find out. This guy made the mistake of saying bring it on. When his order arrived, his eyes went wide, but he had dug his own grave. He went to work, making it about halfway through the tray before you could obviously tell he had gone a bite too far. He ran out the door and a fisherman soon reported the bigmouth was out on the grass shoulder that separated 1A from the pier road purging, or as fishermen say, blowing lunch. He came back and Ute taunted him, telling him how much fun they would have together today. He ate some more, but finally rushed out never to be seen again. Meanwhile, the rest of us had enjoyed a good breakfast and live entertainment, and the time was just approaching six o'clock.

The council raised the cod-trip limit twice in 2000, first to one hundred pounds and then to two hundred pounds. The year passed quickly, and Daniel and I caught a large amount of whiting during the summer. Tom was doing well on whiting as well, but Red had stuck with groundfish because whiting is just too much work without a crew. The NEFMC's research steering committee made good progress on setting up the bid process for cooperative research and a proposal I had made with Dr. Hunt Howell from UNH for cod tagging of cod in the closed area was funded to begin in 2001. I looked forward to tagging cod for two reasons: first, because I hoped it would begin to stimulate discussions on stock boundaries; and second, because it would provide much-needed work that would save me from fishing off Boston in the spring. Our experimental plan called for half-hour tows with the net set up to avoid bottom fish like flounder, dumping the fish into my flooded fish bin, and transferring the fish to plastic vats with circulating seawater and then to a measuring table where length was measured before the tag was inserted and recorded. With a crew, two graduate students, Dr. Howell, and myself, we could tag up to three hundred fish per hour. In 2001, I would tag over twenty thousand cod! The next step was to retrieve the tags from fish fishermen caught. Most tagging studies achieve a 5-to-10 percent return rate, but we hoped for better because fishermen were actively involved in the project. Almost all the commercial fishermen were excited to help, because they hoped this would change the regulations. Our biggest problem was that with the low trip limits, people were trying to return fish alive and would see the tag as the fish went back into the water. The second problem was that the vast majority of Jeffreys Ledge was closed to commercial fishing and

The author, tagging yellowtail flounder for Dr. Steve Cadrin. David is measuring the fish, 2004.

the recreational fishermen who caught cod often did not provide a location other than "Jeffreys Ledge." The ledge is forty miles long, and we needed some degree of specificity. Nonetheless, our tag return rate ran 15-to-17 percent over three years, depending on how you filtered that data. The fish grew an average of a half-inch to three-quarter inch per month, and they contracted back to spawning grounds yearly and expanded out to about fifty miles from the spawning grounds to feed in the summer and fall. Virtually no fish were recaptured east of Rockland, Maine, in Canadian waters, or on Georges Bank proper, but surprising numbers were caught in the Great South Channel east of Cape Cod and off Rhode Island in the general vicinity of Block Island. This would be the first of many studies of cod I would participate in, which, finally, twenty years later, would radically change views of cod stock boundaries.

In February 2001, I received an invitation to testify for the National Academy of Sciences, which had been commissioned to do a study on the effects on benthic habitat by otter trawls. Since this was the type of net I tow, I felt well-qualified to explain its actual performance and not the portrait portrayed to the public by some ENGOs. Throughout the eight council regions, councils were under constant attack by the ENGOs, especially Oceana, to eliminate dragging. An old, but prominent West Coast scientist had published a paper claiming dragging was like clear cutting the forest. The statement was patently false, and should never have survived the peer-review process, but it was seized on by the ENGOs to demand action. Two things I knew: dragging in deep-water, complex habitat like corral would be very destructive, but it was also impossible. Complex ledge and coral shredded any kind of gear currently in existence. Towing in featureless mud and sand bottom probably had short-term effect, but this was minimal because creatures that lived there had evolved to exist in this dynamic environment. Now, how to prove this second point?

My son was a third-year aerospace engineer major. I contacted Eric to ask if he could calculate the true weight of a trawl door on the sea floor. I knew from experience it had to be very light, because otherwise the trawl door would sink into the mud/water interface and the boat would stop. Think about walking across your front lawn in mud season. That is what the water substrate interface is like. Being an engineer, Eric deals in absolutes, but as a fisherman I deal in estimates. Eric wanted several exact measurements, horsepower curves, door spread, actual land weight of the door, and a host of other items. Several days later, Eric produced three and a half type written pages of mostly equations which produced the weight in pounds per square inch of the trawl door shoes on the substrate. For comparison, he also produced the weight in pounds per square inch of a size ten shoe on land for a person of his weight, which was ten times the weight of the trawl door in the ocean!

The hearing was in Boston, and Eric attended after classes. Each presenter was given five minutes to present a summary of their findings but could submit as much written testimony as necessary to back up their points. I decided to skip all the math and walk through the findings in language lay people could understand, ending with the weights on land and in the water to make my point emphatic. The woman who was chairing the panel did not suffer fools lightly. The representative from Oceana went just ahead of me and was midway through his we've got Ted Danson speech when he was abruptly shut down. "Do you have any evidence of a scientific nature to present?" "No, but . . ." "Then you can return to your seat and stop wasting this committee's time!" I was next and launched into my presentation. When I finished, the woman looked at me and smiled, stating this committee has gone through presentations all day that provided very little hard science, and a fisherman had produced the best work of the day. She invited Eric and me to stay and meet the committee. The final report would mostly echo my statements on complex and soft bottom. The language in Magnuson would remain . . . to the extent practicable and not become an absolute ban.

In late summer 2001, the first of a seemingly endless group of commissions was formed by the Pew Charitable Trusts, heirs to the Sun Oil Company fortune, to study the oceans and ocean governance. Pew funded numerous ENGOs that were wreaking havoc with fisheries, including Oceana, and fishermen assumed the Pew Ocean Commission would deliver deeply biased reports. Pew had openly endorsed massive, closed areas as a be-all-to-end-all solution to ocean governance. I believed that closed areas, while they had possible uses in tropical zones, had little practical use in temperate zones where fish moved based on bottom water temperature. Oceana was hung up on habitat protection, but again, I believed you did not need habitat closures; you simply needed to make sure no harmful gear was used in sensitive habitats. At any rate, I went and testified with a laundry list of issues that did need addressing and explained why some ideas that were under consideration would not yield the expected results. Shortly after that testimony, the world changed radically.

September 11 dawned as one of the best days of the best month of the year for weather in New England. I was fishing for whiting eight miles southeast of Hampton. There had been some tuna caught in the area where we were towing, and many tuna boats were anchored in the depth I was trying to fish whiting. Many of these boats put out way too much anchor line, but technically because they were anchored, the boats had the right of way. If you got too close to an anchor, most would call and tell you where the anchor was located, so I had three radios scanning a few VHF frequencies. Just before 9:00 a.m., there was a bunch of chatter about a plane hitting a building. I did not think much of it, figuring a small plane had crashed into a hanger somewhere. Suddenly every frequency came alive. A number of these guys watched television all day to pass the time and some of the comments were ominous. Cellular phones were still relatively new on boats, and I had just recently installed one to inform our manager of my whiting catch. I called my wife, told her what I had heard, asked her to turn on the television and call me back. Then I called the manager to see if he knew anything. A few minutes later, Ellen called back in tears; she had just seen a commercial jet fly into the World Trade Center and the other tower was on fire from a previous plane. I looked up as a steady stream of contrails passed overhead towards Boston and points south. A few minutes later, the manager called back. He had a number of tuna being shipped to Japan hung up in Boston. He asked me to spread the word to the tuna boats to return home as no fish would be shipped. I was told to put as much of my fish as possible into bait and return home when I hauled back.

I put an open call out to the tuna fleet, which was met with some derisive comments, and turned out to the deep water to tow for home. Ellen called again to report another plane had been crashed into the Pentagon and that United States airspace was being shut down. I called my bait customers to get rid of my catch and watched in amazement as the contrails began to lower in altitude as the line of planes diverted to Logan. But what if these planes had crazies on board? I called Ellen again to have her call Eric and tell him to stay out of the high-rise district in Boston and away from BU's two high-rise dorms. We would figure out how to get him out of Boston. I woke up my crew to haul up the net. Clearly the country was under attack, and we should return to port. By the time we were done, two things were noticeable: there was not a contrail in the sky, and there was an F-15 flying loops from the Seabrook nuclear plant to the Portsmouth Naval Shipyard. In all my years, this was a sight I never expected to see.

Fishing, as did many other parts of American life, pretty much halted for the next several weeks. However, doing anything else felt irrelevant. Our family had no one directly impacted, but Eric's dream of becoming a commercial airline pilot after graduation ended. The airlines would not need new pilots for years. The council met less than two weeks after September 11, but the meeting seemed surreal. I went more for something to do than anything else. After the customary prayers, people started arguing about all the stupid stuff they always argued about like nothing had happened. I was appalled. I realize people were being urged to resume living their lives and the government was trying to show leadership, but really? The ENGOs did not waste any time; they sued yet again over cod.

CHAPTER 24

If You Are Not at the Table, You Are on the Menu

Daniel passed his driver's license test in the fall of 2001 and immediately wished to go car shopping. Most young drivers start with a used car, often affectionately referred to as a beater, but not Dan. He knew exactly what he wanted, a new Ford Mustang. I took him to the dealership where I bought our vehicles and introduced him to the salesman. I am sure he assumed Daniel was there for a used car and was more than a little surprised when Dan asked him how many new Mustangs were on the lot. Dan went out to look over the selection while I chatted with the salesman. First, he asked about financing, and I said no, Dan will pay with a check. Then he explained about contracts for minors, which I already understood. I would have to buy the car and transfer it to Dan in a private sale for a dollar. Finally, the salesman had to ask how Daniel could afford the car. I explained Dan had worked for me every day he was not in school since he was fourteen. He saved all his money for a car and for his share of college tuition. We had a deal: Daniel could pick the school, and if he had the grades for acceptance, Ellen and I would pay half the cost and he would pay or borrow the rest. I wanted to make sure Dan had skin in the game, just like Eric. Dan soon returned and wanted to test drive a red 2001 Mustang. We went for a ride and returned to close the deal. I sat back and let Dan speak for himself because I wanted him to learn how to make a big purchase. Dan started by asking the price and followed with a question about whether there was a discount for buying a 2001 model when the 2002 were just hitting the street. The salesman flushed slightly and said something could be arranged. Next, he asked if there was a cash discount as he would be paying with a certified check. A deal was consummated at thirty-five hundred dollars below the sticker price. Daniel drove home the proud owner of a new car. The car would look like the day he drove it off the lot until he traded it in some twelve years later.

Daniel was likely to be valedictorian of his class, so Ellen was somewhat alarmed when the guidance department called requesting a meeting. Daniel had written his college essay, which the department had reviewed. His essay was about the family tribulations surrounding the fishing regulations of his high school years. The counselor was concerned he had either made the whole story up or embellished it to gain college admission. Daniel is a beautiful writer and the story brought tears to our eyes, both of frustration and joy. I assured the counselor the story was 100 percent accurate, but that even Ellen and I had no idea the trauma Dan had suffered as he internalized everything. Dan would be accepted at the College of William and Mary, his first choice.

On an early winter day, Red stopped me in the coop parking lot to talk. He had been talking with Bob, the manager, about taking a job with the coop as Bob's second in command. Red was fed up with the cod limits, vessel monitoring requirements, logbooks, and now a requirement to carry an observer if requested. Red fished alone and went when the weather and the spirit moved him. If you had to carry an observer, you could not leave whenever you felt like it. For Red, that was the final straw. Although I had seen many local boats quit the fishery in the previous several years, this was the first member of my close fishing friends to throw in the towel. My friend Erik, who was still on the council, had moved in the other direction. He had bought a new, bigger gillnetter so he could fish monkfish on the far side of the western Gulf of Maine closed area. Monkfish gillnets had extra-large twine, which was exempt from observers. You only hauled the gear every three to seven days, as the monkfish stayed alive for quite a while. This fit well with the number of days Erik missed to attend council meetings. The Portsmouth Cooperative would fold in the spring of 2002. You simply could not land enough fish to pay the bills. Erik was the president when the financial stresses became obvious, and he had to spend a great deal of time negotiating an orderly closing of the business.

I put in another bid to study cod with Dr. Howell at UNH. This time, we would tag fish in the western Gulf of Maine closed area. This was one of the most difficult projects I attempted. Most of the closed area was untowable, and because it was closed, the lobster fishery had set thousands of traps there. Also, fishermen fishing the adjacent open areas took all their derelict traps over the line and dumped them off. Every day was a struggle, but we still were able to tag a respectable number of fish. As expected, many returns came from the recreational fishery fishing the closed area, but a surprising number came from

Ipswich Bay in the spring and Massachusetts Bay in the winter. Apparently, both groups of spawning fish summered together on Jeffreys feeding on herring and krill before returning to their respective spawning areas.

In the spring of 2002, I was approached by a council member from Maine and a professor from UNH about being part of a cooperative research course they hoped to offer. The course was titled the Marine Resource Education Project (MREP) and would have two, three-day modules, one on science and one on management. My job was to be the moderator/translator for the science module. The course would select twenty fishermen and five "others" (non-fishermen) to take each module. Everyone, both lecturers and participants, would be fed and housed at the UNH hotel school. Fishermen would be paid a stipend for lost fishing time. Everyone had noticed that fishermen clearly did not understand the science and often became involved at the last minute in council decisions long after they could meaningfully affect the outcome. Lecturers were selected based, first, on availability, and, second, on communication skills. Could they listen as well as talk? The first presentation was a trial run and proved very illuminating. Fishermen, scientists, and managers speak three different languages, all loaded with idioms and acronyms. As translator, I had to rephrase statements often discreetly into the language of the speaker. I found all the presenters really wanted to be understood but could not always answer questions using simplified science terminology. Fishermen likewise asked questions based on their fine-scale knowledge of fish distribution using local names of ledges or grounds that scientists had no knowledge of. The non-fishermen had no knowledge of gear types or terms, how gear worked, or what gear caught and why it caught what it caught. I felt like the only translator at a United Nations General Assembly meeting. We quickly learned that presenters would need to have hands-on exercises to make their points and be more visual. Instead of talking about how fish are aged using scale samples and otoliths, bring a fish and dissect out the otoliths and show both the scales and otoliths under a microscope. While you were at it, identify the organs of the dissected fish and what their function is. Fishermen pull guts out of fish hundreds of times a day for decades, but one would be surprised how many have no idea what is in that handful of offal. Having people dine together also proved successful, as everyone got to see everyone else as someone like them who liked beer or sports or whatever topics were brought up. The course proved very successful, and I moderated the science module until 2010. The course is now given in different regions throughout the country and has helped numerous scientists and fishermen forge bonds to conduct cooperative research.

In May of 2002, Judge Gladys Kessler ruled mostly in favor of the Conservation Law Foundation (CLF), and ordered a negotiated settlement which would become the basis for Amendment 13 that the council would pass in 2003. Basically CLF, state directors, and some lawyers representing some of the big players would be sent to a hotel, not be allowed to contact the outside world, and negotiate a settlement that the judge would order the government, through the council, to impose. An adage goes, if you are not at the table, you will be on the menu. Well, our association and several from Gloucester were not at the table. Worse, my state director never consulted with any of us as the judge instructed, but others phoned people within their states. It would come as no surprise then, that the most onerous restrictions in the settlement were imposed on coastal boats from New Hampshire and northern Massachusetts. This was the last straw. A few fishermen had good ideas about fixing our endemic problems, but no one ever listened because the politically entrenched, the 10 percent who now caught 90 percent of the fish, controlled the process. The mayor of Gloucester invited all fishermen, of all gear types, from all regions to attend a meeting to form a new organization. The goal would be innovative solutions to groundfish fishery problems that worked for all fishermen and all gear types in all regions. You would be expected to contribute 2 percent of your groundfish revenue to hire an executive director and a Washington lobbyist to work with Congress to correct some of the outstanding regulatory issues with the current iteration of Magnuson. This group became the Northeast Seafood Coalition, and in my opinion, it has been the best thing to happen to fishery management in my lifetime.

Eric graduated from Boston University in May with honors in aerospace engineering. He returned to Hampton to run a six-passenger charter boat while finishing his commercial pilot licensing requirements. Daniel graduated high school shortly thereafter as class valedictorian. He had chosen the College of William and Mary in Virginia for several reasons. He like the academic rigor, and the school was in the heart of many historic battlefields. After fishing, the Civil War was Daniel's favorite interest. Daniel would be leaving for Virginia after the third week of August, which meant the search was on for a replacement.

The world had changed and the men who worked on boats as professional crew were pretty much no longer around. I placed my first classified ad ever. This was an eye-opening experience. I had one applicant who planned to walk from Rochester, New Hampshire, about forty-five miles away, every day. Several others wanted to live on the boat. Still others only wanted to work when it was convenient for them. Of the several that sounded like their seabags were at least partially packed, we began an on-the-job interview process. The job interview went something like this. Do you have a pulse? Yes. Is it somewhat regular? Yes. The reader would be surprised how many failed those two simple questions. Next, can you get here at 4:30 a.m.? Yes. Do you have a phone or can you give me a phone number where you live? Maybe. The first three we tried out were no shows. I would call the phone number and it was either disconnected or the voice on the other end never heard of the person I was looking for. The joke around the pier was that the aliens must have abducted them, because they vanished without a trace. Finally, a retired US Army Ranger applied. I had high hopes, but the man got desperately seasick and wisely decided the job was not for him. Shortly after, a man applied who had some crew experience. I checked him out and found a problem with alcohol (no driver's license), but he lived in town and offered me a key to his apartment in case he did not answer the door. If he made it to his apartment at night, at least I knew where to find him.

In mid-August, I was requested to testify for the US Commission on Ocean Policy at Faneuil Hall in Boston. Congress had formed the commission out of fear the Pew Commission might issue a biased report and box Congress into action. The committee was chaired by the former chief of naval operations, Admiral Watkins. The panelists were a who's who of ocean expertise. The charge was much larger than the Pew Commission, which was mostly fish and habitat related, being told to investigate all aspects of ocean policy. My topic was fishery policy and how to strengthen it. Each speaker was given ten minutes, no exceptions. I decided on bullet points and listed twenty items in need of change or improvement. At Ellen's suggestion, I read my statement in front of a mirror: it ran nine minutes and twenty seconds, if I kept my cool. I decided to contact the Northeast Seafood Coalition's lobbyist in Washington and ask him to review my statement. Glenn had an encyclopedic memory of the Magnuson Act as well as several other legislative acts that controlled fisheries. Glenn agreed I had covered as much as you could cover in ten minutes and offered some phrasing changes to make some statements clearer. He also suggested I include an explanatory paragraph on each bullet point in my written testimony. So, on the appointed day, I put on the only suit jacket that I had owned in my life, and a tie with an American flag emblazoned with "Don't tread on me" with the serpent which the navy had been authorized to fly second staff since 9/11, and headed for Boston with Ellen. I figured if it was good enough for John Paul Jones, then hopefully the admiral would appreciate my knowledge of history.

The panel was made up of about sixteen people and most names were readily known at the time. I was last on the list of presenters again. This was useful, as you got to understand what resonated and what crashed and burned. This was the first time I testified where the chair controlled the microphone. I had watched several people get shut down at precisely ten minutes. The panelist ahead of me was one of those individuals. The chair of the New England Fishery Management Council used half his allotted time telling the panel who he was. At this level of government nobody cares. He had barely launched into his meat and potatoes when the microphone shut off. The admiral told him politely but firmly to return to his seat in the audience. He looked stunned and started to argue, but that only got him a look that could melt glaciers. He wisely returned to his chair. I remember it was hot and I was roasting, and the room had not cooled down any after the last exchange. I launched into my presentation and finished with time to spare. The admiral seemed pleased and invited panelists to ask questions. With those behind me, the hearing was adjourned. Ellen was beaming when I started to return to my audience chair. The admiral motioned me back, shook my hand, and handed me a large coin. He told me it was a challenge coin awarded for outstanding service. I did not fully understand the significance, but Ellen did, as she had an uncle who survived Pearl Harbor. The final report would be issued in 2004; parts of the section on fishery governance are nearly verbatim quotes from my testimony.

Ellen set Daniel up at William and Mary and came home. We were officially empty nesters. Now there was just Ellen and me and Daniel's pet birds, a sun conure named Horus and a severe macaw named Skye. The conure was extremely put out for about three weeks and spent all day looking for Daniel. Finally, I guess, the bird decided Dan was gone and adopted me. It would fly to me and crawl in my shirt, turn around so its feet pressed against the fabric, stick its head out above the collar and go to sleep. Horus only liked the immediate family. It would attack anyone who walked in unannounced. We put up a sign: "Warning: attack bird!" but I had a crewman who thought the sign was a joke. He walked in one day and Horus attacked him, biting his neck and flying off to turn around and attack again. He was going for his jugular vein and scoring. The crewman was screaming, curled up on the floor, and the bird just kept at him while I tried to wrap Horus in a towel. Finally, I caught and caged Horus, but not before he scored seven direct hits. The crewman looked pathetic, muttering something about it being just a bird. I said, "Well, 'just a bird' did a pretty good job on just a man!" He never failed to knock again.

As 2003 rolled around, I was searching for more research to become involved in. I heard from a professor in New Bedford, Steve Cadrin, who was putting together a comprehensive plan to tag yellowtail flounder, another one of our problem species. Specifically, he wanted hundreds of fish tagged in Ipswich Bay, which was considered the northern end of the species range. Most of the work would be done on Georges Bank and in southern New England by other boats, but the project proposed to tag fish over multiple years to account for interannual variability, which is a fancy way of saying changes in movement patterns from year to year. Tagging would be during spawning season when it was assumed the fish returned yearly. That would be May and June when Ipswich Bay was closed. I signed up. Work would begin in May.

CHAPTER 25

Death of an American Original

By mid-May, my son Eric had completed his commercial plane license requirements and went to New Jersey to check out a pilot job. Daniel returned home for the summer about the time Eric left. Daniel had no declared major at William and Mary and took classes in various subjects that interested him. He could stay undeclared for two years, but then had to pick a major or his advisor would pick one for him.

We went right to work on the yellowtail tagging project. The *Ellen Diane* was loaded with big tubs approximately four-feet cubed. The fish would be transferred from the fish bin to the tubs, which were filled with circulating sea water, and then be captured with a dip net for tagging. On the tagging board, the fish were given a numerical score for overall condition and candled for a parasite called lymphocystis. Then they were measured, tagged, and returned to the sea. Lymphocystis was a parasite that was showing up in increasing numbers of yellowtail and appeared temperature-dependent. The parasite was thought to increase natural mortality by gradually weakening older fish until they finally died. Surveys catching yellowtails showed very few fish over age eight, which, of course, was blamed on overfishing. I had never heard of this parasite, so this information was all very interesting to me.

This is how cooperative research shone, fishermen and scientists learning from each other. What I did know was our tagging methodology was slow and inefficient. Yellowtails are flounder, which live on the bottom. Our bottom was under four feet of sloshing seawater. Catching the fish with a dip net took longer than dragging them up from the bottom. That night I went to Walmart and purchased two kiddy pools about four feet in diameter and a foot deep for ten bucks a piece. The next day, we used the pools. People could wear a rubber glove and pick the fish out of the pool. We probably tagged ten times as many fish on day two as were tagged on day one. Kiddy pools became the new norm in flounder studies, a low-tech solution to a high-tech problem.

The rest of the fishing year was uneventful for Daniel and me, but Eric's job in New Jersey did not pan out. Ellen and I had a tradition of meeting two couples at one couple's lake house for a cookout in late August. Both men had worked for me as crew while attending college in the late 1970s and early 1980s. We had remained friends since. One of the men worked for Boston Transportation, the largest tugboat company in Boston. As we discussed Eric's difficulties in locating a pilot's job, Bill suggested Eric apply for a job with his company as an engineer. Eric could still fly on his days off and this would be solid employment. Eric was deeply disillusioned with the chances of gainful employment in the airline industry and took a job on the tugs. Eric soon gained the nickname MacGyver, as he seemed able to temporarily repair anything with his Swiss Army knife and duct tape.

Just before Christmas 2003, the council released the final version of Amendment 13. Scrooge apparently had not yet been visited by the ghosts of Christmas. More months of closures were added, days at sea were cut by 40 percent, but hey, now you could lease days back from your neighbor who did not fish and the cod-trip limit was increased to eight hundred pounds. Fishery management, scientific orthodoxy, and the ENGOs still held to a binary equation: reduce fishing mortality far enough and stocks would rebuild. I have never believed that equation, because it assumes that simple probability is real, i.e., more adult fish yield more juvenile fish. Throughout my lifetime, I have seen small-year classes produce high recruitment and large-year classes produce next to nothing. Why? Because Mother Nature deals the cards. Recruitment is a function of physical oceanography, having the fish spawn in the right place at the right time at the right temperature. Then the eggs must stay in waters, then when they hatch and the larva settle to the bottom, the right plankton has bloomed at just the right time for the larva to survive. The odds are still daunting, an adult cod may produce 3 million eggs of which maybe two survive to reproduce. If that number doubles to four you have a major recruitment event, if it drops to zero you get nothing. Why pursue a fantasy? Well, absent that notion, one would be forced to admit that all you can manage is people and that nature will respond on its own timeline and not the one put into Magnuson by Congress. Einstein is purported to have said trying the same thing over and over again and expecting a different result is the definition of insanity. Luckily, he was not around to observe modern fishery management.

The shrimp season lasted various time periods every year, up to a maximum of six months. The longest seasons ran from December 1 to May 31. During the early 2000s, seasons ranged from four to six months. During 2003, the season ran through April, which meant we were fishing in deep water on the western side of Jeffreys Ledge.

By then my father lived in an assisted living facility in Hampton, where he kept the staff on their toes. We had finally gotten him into the facility after he started using stock certificates to light the wood stove in his Exeter home. He also was dumping all his prescription medicines in a candy bowl and would scoop up a handful if he was feeling a little off. After one of many trips to the emergency room, which was less than a block away, the attending physician was about to release him yet again, when I intervened and asked the physician to ask him a few questions to test his mental acuity. My father was hallucinating and making statements that made no sense but could pull his act together long enough to pass a cursory examination. After a few softball questions the doctor asked him who the president was. Response: "That asshole!" "Well, I know he is, but what is his name?" "I do not know; I'm tired." "Why are you tired?" "The damn Pakistanis in my basement." "Do you rent your basement?" "No. They snuck in through the basement door; there are hundreds, and they make noise all night." The doctor replied, "Really?" and motioned to a nurse to take Mr. Goethel back to his room.

My father insisted on having his car at the facility even though he was not allowed to leave. The staff parked the car where he could see it, and despite having confiscated four sets of keys, he kept finding others. He would then round up several of the female residents and they would escape on a road trip. This happened three separate times with the car returning with increased damage every time. The last one must have been a hell of a ride as one wheel was being driven on the rim and two others were flat. I checked with the police, and no one reported any damage, but enough was enough. I pulled the wire from the coil to the distributor and took it home. If he found another key, the worst thing he could kill would be the battery. So, it was not surprising when I received a call from Ellen saying my father had a major stroke and the hospital would not enforce his living will until I arrived. I was three hours from the pier, and despite a call to the hospital, explaining the situation, my father was given lifesaving intervention

I wish I had caught this on my rod and reel! Jack Goethel aboard the *Ellen Diane*, late 1980s.

despite his express wishes. He would hang on for fourteen days, hooked to virtually every machine the hospital owned, before his heart finally gave out. This was a sad and unfortunate end to an American original.

Amendment 13 would go into effect May 1, 2004, and it contained an enormous number of changes. For example, boats allocated eighty-eight days at sea but which never fished would have their days reclassified as C days. They would not be allowed to use these days. Boats who fished, say eighty days, would have those reclassified as A days and be allowed to fish 55 percent of them. The remainder would become B days, which could be used in several special-access programs with special gear. The cod-trip limit would increase to eight hundred pounds and a leasing program was created to lease A days from your neighbor if he had a similar-sized boat. The amendment stated it would end overfishing on all stocks and rebuild those that needed rebuilding within allotted time frames. Before the ink was dry, the ENGOs would file yet another lawsuit. Daniel returned from William and Mary with a perfect GPA. His advisor had decided Dan would declare a physics major and math minor. Why? Because it was the hardest undergraduate work they had. Daniel liked the math but was not impressed with physics. Nonetheless, he would complete the required course load. During the yellowtail tagging, Steve and his graduate student would try to engage Dan in conversation, and when he did not respond, they assumed he did not like them. This was not true; Dan just did not talk. Some days when it was just the two of us, he might not utter a sentence all day. He reveled in the solitude that the vastness of the ocean provided.

I had applied for a seat on the council, as my friend Erik was termed out. Under Magnuson, you could only serve three consecutive, three-year terms. Erik was happy to depart, as he was tired, frustrated, demoralized, and burned out by the incessant attempts to satisfy both the law and the ENGOs. To be appointed, you had to survive a two-step process. First, the governor had to provide NMFS with a minimum of three qualified individuals. The governor could rank the people in order of preference or just provide the names. Next, the process moved to NMFS, where the upper echelon of the bureaucracy vetted the names, took comment from the public, including congress, and tried to provide some balance among competing interests. The process was political, and by trying to satisfy everyone, you seldom satisfied anyone. Somehow, my number came out of the hat, and I was appointed in June to start a term in August 2004.

Shortly after that announcement, I was contacted by the editor of the *National Fisherman* magazine. They were planning on nominating me for their highliner award and wanted to send a reporter for an interview. The *National Fisherman* was the only countrywide fishing paper, and this award was a major achievement. The paper nominated three people from the East and Gulf Coast one year and the West Coast, Pacific Islands, and Alaska the next. One of my fellow nominees would be Jimmy Ruhle, who also served on the research steering committee and had just been appointed to the Mid-Atlantic Council from North

David making a speech? Where is his foul-weather gear? A fish out of water! *National Fisherman* Highliner Dinner and Award, 2004.

Carolina. The term "highliner" originally referred to captains who caught the most fish but had changed over the years to mean both catching fish and leading the industry forward. Cooperative research was widely viewed as producing useful results in a cost-effective manner. The award was a recognition of leadership as much as it was about our fishing talents.

I was sworn in at the September council meeting.

Proud of each other. Dad, at the top of his game, and son, aerospace engineer and soon-to-be tugboat captain! Highliner Award, David and Eric Goethel, 2004.

Being on the council did not seem difficult to me, as I had participated in the process for years. The process was not the problem, but politics was. I was not political in that I would not trade favors or undermine someone to win. What you saw was what you got. I made fishermen mad, scientists mad, and managers upset, and on a rare day scored a hat trick as everyone was mad. Even so, I got credit for reading the council binder cover to cover and having my points meticulously outlined to thwart opposition. I knew *Robert's Rules* and often could use parliamentary procedure to my advantage. The council is a voting body by a majority of members present. Most votes are hand votes, and humans being humans, many want to be on the winning side. Consequently, controversial votes have a lot of looking around before some hands go up. Because I felt people should be on the record on controversial votes and because I thought the public should know how council members voted, I frequently exercised the right to call for a roll call. In this case, the computer would shuffle the names and people would be called randomly for their vote. I always appreciated the deer-in-the-headlights look some of the political types got when they had to go on the record without knowing the outcome. On average, I usually picked up two votes, often going from a nine-to-seven loss to a seven-to-nine win. I was not popular with some council members or with many lobbyists, but I was there to represent the fish, who did not have a vote. Most of the hard council work is done in committee meetings. You usually serve on three to four committees. Most of my council years, I was on groundfish and small-mesh multispecies (whiting), and my entire tenure I was chair of the research steering committee.

Groundfish was a perennial pain. You would spend hours crafting a solution only to have it torn apart by the full council. Most of the council did not care about whiting or research steering, and this indifference allowed you to get a lot more constructive work done. In addition to making funding recommendations, research steering oversaw grants and contracts and served as peer reviewers for final reports. Peer review is one of science's most important functions. The review is supposed to make sure the results are verifiable and repeatable and that the report sticks to facts and does not stray into opinions. There is also a code of conduct, rules such as you cannot review your own work or that of a relative. A peer review is not an atta-boy for the scientist in the adjoining cubicle. I had watched lax peer review create ongoing problems in fishery science and was determined that would not happen on my watch. Consequently, a half day with the research steering committee was not high on some researchers to-do list.

We also oversaw budgets, and I was constantly amazed at what some people considered essential equipment and overhead. The first few years, we zeroed out new pickup trucks, office computers, and equipment most functioning labs should contain. After a while, researchers realized their budget would really be scrutinized, and so that type of stuff stopped. We also had to verify progress reports. Most researchers were very conscientious about following their research proposal, but there was always one who thought that they could use the money

The full New England Fishery Management Council. David Goethel, *second from the left, front row, 2008.*

as they saw fit. One notable incident occurred when I called for a vote to suspend payments to a very prominent university's professor for not following the terms of his research proposal. I was soon on the phone with the council's executive director, who had a very angry dean on another line. I explained the vote; he backed my call, and I was off that Christmas-card list. Another incident required me to referee a dispute between a researcher, a council staff member, and a science center scientist. I was dispatched to the southern New England coast by the executive director to assess the situation, assign blame, and craft a solution, if possible, or at least get everyone to play nice in the sandbox. Science has some very strong personalities and attracts people who have not always been trained in social graces. In this case, there were just inherent personality conflicts that could only be solved by assigning people to other projects. I was beginning to wonder if I was a kindergarten cop or the council's pit bull, but either way, the problem was solved, which was all I cared about.

I think people involved in the fishery management process sometimes forget that fishing, in the best of circumstances, is dangerous work. I dealt with many people both in the regional office and the science center as well as council staff, who seldom, if ever, went to sea. Writers have tried to describe the perils of storms and the misery of seasickness and extreme fatigue from twenty-hour days, but if you have not lived the life, you really cannot talk about it: words simply do not do life at sea justice. Gordon Lightfoot perhaps came closest to the terror that occurs when he wrote in *The Wreck of The Edmond Fitzgerald* about the way time seems to slow down in times of danger. The reason I write this is that one can easily forget that fishery management puts men in impossible situations to feed the American public. Two men who did have plenty of experience at sea were council executive director, Paul Howard, and chief staffer for groundfish, Tom Nies. Both were retired from the Coast Guard and had spent years at sea on Coast Guard cutters interacting with fishermen. Sometimes you could watch the frustration creep across their faces when a career bureaucrat made a statement that no one who spent time at sea would ever make. Tragedy at sea usually occurs not from one catastrophic failure, but rather from a series of small failures that occur in rapid succession. This was a point made often to me in my early fishing career. The chances of tragedy at sea increase exponentially when you push a boat beyond the range it was built for. This was beginning to happen now as small boat fishermen tried to fish outside the closed areas. Sinkings and near sinkings increased as desperate people took big risks. I used to become infuriated when an ENGO said: "Well, they do not have to go out there, they could just stay home." There is no unemployment insurance or benefits for self-employed people; you do not fish, you do not eat.

The ENGOs largely lost their lawsuit over Amendment 13. A federal judge ruled that the amendment was a good-faith attempt to implement the negotiated settlement that the ENGOs themselves had agreed to and the parts NMFS had rejected were based on sound legal reasoning. Further, she ruled the phrase, to the extent practicable, had a degree of reasonableness associated with it, not absolutism as the ENGOs claimed. Now maybe we could get down to fixing our problems, but no, the ENGOs went to work on Congress. Magnuson was coming up for reauthorization soon, and there was serious lobbying to be done. In return for nonprofit status, ENGOs are allowed to "educate" but not to "lobby" Congress. Like many rules, this one is never enforced. Education requires telling the entire story; lobbying requires telling what you want the listener to hear. Many parts of Magnuson needed reworking, like goals for keeping all stocks at maximum sustainable yield (MSY) simultaneously, which is ecologically impossible, or mandatory rebuilding periods not to exceed ten years, which likewise was ordering nature to do something that might not happen even absent fishing. All stocks at MSY fails to take account of predator-prey relationships. When the foxes are at high abundance, the rabbit population will be reduced. Nature modifies spawner-recruit relationships, and absent continuous favorable recruitment events, many stocks will take longer to increase. While groups like the Northeast Seafood Coalition submitted white papers for congressional staffers to read, the ENGOs continued a pitched battle for keeping status quo language as well as adding hard quotas to Magnuson. This was going to be an interesting battle.

CHAPTER 26

The Fishy Gene

Both fishing and management seemed to fall into predictable patterns by 2005. No one could argue that there were more groundfish of all species everywhere, even if you could barely fish them. The ENGOs had stopped the lawsuit mill, at least for the moment, and the council turned more of its limited attention to other managed species. Everyone spent all their time castigating council management of groundfish, but, taken on the whole, management of most species was doing very well. Scallops were the biggest success story, with yields no one thought possible on a continuing basis. Even groundfish was showing improvement. No one recognized the issue at the time, but the waters off New England began to warm. Shrimp populations were the first to show signs of reacting to warming water. The Gulf of Maine was the southernmost range of northern shrimp, and without very cold water near shore in February, shrimp spawning success dropped dramatically. Shrimp seasons were shortened, but it had little effect on getting bigger year classes of shrimp.

Daniel was in the early stages of scoping out graduate schools during the yellowtail tagging in May. He did not want to do graduate work in physics but was looking for something math-related. He still never talked on the boat but had struck up an email correspondence with Dr. Cadrin, who taught at the School for Marine Science and Technology (SMAST) in New Bedford. SMAST was trying to rapidly build a world-class graduate-school program in fisheries and was recruiting the best and the brightest for its program. Daniel did not like to talk, but he loved to write and would converse via email. Dan did not have formal training in biology but had informal training in invertebrates from his mother and vertebrates from me. Dan was considering doing graduate work in stock assessment. There was just one little problem: Daniel loved to fish.

Dan had what I call the fishy gene. All the talented fishermen, both commercial and recreational, have the fishy gene. Call it thinking like a fish or some sixth sense that lets them understand fish movement, I do not know. I can only describe what I feel. I can sense fish movement and calculate, based on a few factors, where fish will be and what they will be eating at any given time. Daniel demonstrated this talent daily with striped bass. When we were whiting fishing, on the last tow, Dan would drift out a bait and immediately hook these enormous stripers. Commercial boats cannot keep striped bass in New Hampshire, so he would just wind them up and let them go. He would do the same thing at the coop dock while we unloaded. The dock always had several people trying to catch striped bass. Some of these people would fish all summer and maybe get one or two fish. Daniel would toss his line in while hooking up racks of boxes and be playing a fish within minutes. With great fanfare, he would unhook the fish and let them swim away. People on the dock would beg him to give them the fish, but he would always solemnly say, "Sorry, commercial boats cannot keep striped bass." Secretly though, I could see he reveled in the notoriety. At some point, Dan would have to come to grips with either taking over the boat or pursuing a land-based career.

My cod work was largely concluded, with peer-reviewed papers published, so I began to cast about for some new research. A new data-collection program called the study fleet was trying to get up and running, but not receiving much enthusiasm from fishermen. The goal was to have fishermen collect fine-scale, catch-and-discard data by tow, and the information would be used to enhance assessments. Fishermen were worried that providing data on where they caught fish would lead to more closures. They had a point, as most of the best fishing grounds were now closed when fish were present. My feeling was the government already knew where you fished. We were all required to have Vessel Monitoring Systems (VMS), a kind of ankle bracelet for the boat which transmitted your location every half hour to enforcement. I was requested to sign up to lead by example. Initially, this was a lot of work, as the computer system was still in the development phase, but one of the perks was a temperature/depth sensor which would download on the computer when the trawl door surfaced. I already knew the depth and what I caught, but now I had the last piece of the puzzle: the bottom water temperature.

Through my translator/moderator role in MREP, I had learned just how little the people who wrote the laws and the rules understood about fishing gear. So, I worked with people to design a fishing gear workshop for non-fishermen.

We would have part lecture and part hands on with gear in a parking lot, but the final piece would be taking people out and setting, towing, and hauling up the gear, to be followed by a quick biology lesson. We would demonstrate both trawling and gillnet gear using two gillnet boats and two trawlers. Initially, we limited participants to NMFS staff and congressional staffers, but we later expanded to the Coast Guard, state fisheries, and ENGOs. Part of the lectures were discussions on what fish could sense, how deeply light penetrated the ocean, and where fish would attempt to escape the net. These talks were given by gear specialists at UNH. The program was very well-received and highlighted the value of cooperative research to two very important groups of people. One of the researchers and I began discussions on further bycatch reduction in the shrimp fishery as an outgrowth of his talk. This would lead to the development of the topless shrimp trawl, a purposely named double entendre to get fishermen's attention.

In March of 2005, we went to Newfoundland to test our experimental design in Memorial University's flume tank. Basically, you build a scale model of the net and place it in the tank, which has water flowing from one end to the other to simulate towing. The goal is to see if the net performs as expected, has stable geometry, and the right number of floats, and whether the sweep maintains contact with the bottom. You can take detailed measurements and pictures and do more in a couple days than you can do weeks at sea by trial and error. Our design clearly showed deficiencies. The twine toward the mouth of the net formed a semi-circle and bowed down over the sweep, while at the center, the sweep did not maintain bottom contact. Adding flotation lifted the twine but made the bottom contact even worse. After discussion with the man who had built the scale net and ran the tank, we all agreed that a taper of the twine at the mouth of the net called the lower wing had to be radically changed to remove a lot of twine. We were discouraged, as we only had the tank booked for two days and recutting the taper would take the better part of a day. The man said not to worry, he would stay up all night recutting the taper and we should go to dinner with the dean as planned. The people of Newfoundland are the kindest people I have ever met. This was my third trip to St. John's, and everyone I ever met there went out of their way to welcome strangers and make you feel at home. The next day, the net performed flawlessly, and Memorial drew a detailed net plan that I could transmit to my net builder so we would be ready for the winter shrimp season. In February 2006, we made numerous paired tows comparing the topless trawl to my traditional net. Two items stood out. Bycatch of herring and haddock, two fish that escaped through the top of the net, dropped by almost 90 percent, and, better yet, shrimp catch increased by almost 15 percent. This was the best of both worlds. We published our results and won an award in the smart-gear competition. I continued using the net until the shrimp fishery closed due to declining stocks caused by the warming of the Gulf of Maine in 2014.

The year 2006 was shaping up to be very busy. Sea Grant in California at UC Davis had organized a trip, in conjunction with the ENGO, Environmental Defense Fund (EDF), to study the New Zealand Quota Management System. New Zealand was the darling of several groups then as they had hard quotas with no discard, instant fines for fishery violations, and strict adherence to scientific recommendations. Sea Grant hoped to present the entire story, but EDF hoped to indoctrinate Americans in their goal for future fishery management under the revised version of Magnuson. I went for one reason: I could never afford to travel to New Zealand, and even if I did not like the management system, I would get to see the country. A fellow council member and I quickly got the lay of the land. First, the law was very different there, and many of their management techniques, like instant fines, would be unconstitutional in the US. Second, there was no mandatory rebuilding, and quotas could vary no more than 15 percent per year. Finally, we learned that a mere six corporations controlled 92 percent of the quota in the third-largest exclusive economic zone in the world. Vice presidents of these companies built and raced America's Cup yachts while the captains of boats were referred to as human capital and barely paid a living wage.

Two things helped our fact finding: New Zealanders like

All by myself, unloading an epic load of northern shrimp. The crew didn't show up this morning, but I slept well that night, circa 2006.

beer and my traveling companion was a very big man. Since the companies sent corporate representatives to stare down employees and make sure they spouted the corporate line, my friend would corner that person in a corner of the vessel's stern and engage him in endless questions while I got the real story from the crew. My friend was so big the minder could not get around him. In the pubs where we had dinner, invited fishermen would follow us into the restrooms and unload on how they had been forced out of the fishery and made to sell to the sea lords. The travelers quickly fell into two groups: those who began plotting to corner their fisheries and the rest of us, who dreaded the fight ahead. Still, the scenery was outstanding.

Daniel graduated William and Mary with a nearly perfect GPA. He had been accepted at graduate programs at Duke, Cornell, the University of Rhode Island, and SMAST. Dan was emailing Steve almost every night with questions about where to go. I know Dr. Cadrin wanted Daniel tremendously, but he showed extreme professionalism by pointing out all four programs' strengths and weaknesses. Ultimately, Dan chose Steve's program for several reasons: they had developed a genuine rapport, Daniel could continue working on yellowtail, and SMAST offered Dan a combined master's and PhD, which would shorten his study time. Daniel also received a full scholarship and a stipend to live on.

By now, cod were extremely easy to catch. The enormous 2003-year class of Gulf of Maine cod and Georges Bank haddock were entering the fishery. These two groups of fish were examples of getting the right natural conditions for small-year classes of fish to produce large recruitment events. Dan and I moved to Gloucester for the month of June. We would leave the house at midnight and drive to Gloucester, arriving shortly after 1:00 a.m. I had rented dock space in East Gloucester. Usually, we would be on our way out of the harbor by 1:30, with the ride to Middle Bank taking slightly less than two hours. The cod here were feeding on sand eels, a fish about the size and width of a pencil that burrows in the sand by day but comes out at night to feed. Sand eels are very high in fat and oil, and cod just gorge on them until they are literally hanging out of their mouths. The bottom looked like a carpet of fish. We would set out and literally tow five minutes and haul the net up, usually this yielded between eight and twelve hundred pounds of fish, almost exclusively cod. We would box nine and a half boxes, which when gutted and gilled, would yield our eight-hundred-pound limit. We would then steam back to Gloucester to get inside the VMS line to shut off the day at sea clock, and finish cleaning the fish. There was only one problem: sand eels make the cod soft. They will literally burn through the flesh as they decay in the cod stomachs. Because of this, we had to pull the cod's gills out. There was always a blob of sand eels behind the gill rakers. This helped improve the quality. Still, the dealers paid less money for Middle Bank cod because they were soft, but if you went over towards Boston to catch cod feeding on herring, you still got a lower price because there were plenty of Middle Bank cod to buy. Fishermen never beat the dealers at their own game.

Reauthorization of Magnuson in 2007 was a complete disaster from my perspective. Not only were the biological impossibilities not removed despite numerous letters from people knowledgeable about the subject from many different regions, but mandatory hard quotas were added along with limited-access privilege programs. This was government-speak for limiting Americans access to the fishery while calling fishing a privilege. I wonder how many of these geniuses would consider fishing a privilege when they were boxing shrimp in ten-degree weather in a howling snowstorm. Further, the Environmental Defense Fund had gotten the concept of sectors into the act. Between limited-access privilege programs and sectors, they were giving economists a chance to screw up the fishery. From where I sat, you had now introduced greed, one of the basest human conditions, as a management tool, but economists would say they were removing inefficiencies from the market. What could possibly go wrong? Meanwhile, a little-known woman named Jane Lubchenco, a vice president from EDF, was giving speeches to California venture capitalists urging them to buy up fishing permits from struggling, individually-owned small entities and form large fishing corporations. The sales pitch: "This is not 'dot Com' money, but you will receive very healthy returns on your investments as there are only so many fish and you will own them." There was only one problem with this line of thinking: corporations are interested in return to investors, not sustainable fisheries. The reauthorized act would be signed by the president in the summer of 2007. I had been reappointed to the council in June. Three years of frustration were about to begin.

My friend Tom spent most of the year fishing for silver hake or whiting with the *Midnight Sun*. He had been granted very few days at sea from Amendment

5 and really had no choice. While the *Midnight Sun* was a large boat in Ipswich Bay, it was a small boat on Georges Bank and even smaller in the winter fishery in the canyons on the southern flank of Georges Bank. Tom had been searching for a bigger vessel and found a beautiful southern shrimp boat in excellent condition in the summer of 2008. The *Midnight Sun* was put up for sale and soon had a buyer from outside the country. The boat was run down to Miami to be turned over to the new owner's representatives in late August. The foreign crew was warned a hurricane was approaching, but for whatever reason set out from Miami anyway. A newspaper article from Florida tells the rest of the story. The US Coast Guard received a mayday call from the *Midnight Sun* forty miles south of Great Inagua, Bahamas. The vessel was taking on water in thirty to thirty-five-foot seas and seventy-knot winds. A Jayhawk helicopter was dispatched from Clearwater, Florida, and miraculously rescued the crew, landing them at Providenciales, Turks and Caicos Islands. All four men were saved from the jaws of category 4 Hurricane Ike, but the *Midnight Sun* was never seen again. Tom was sad that his boat met such a tragic end, but his new *Midnight Sun* has proved her worth since, landing millions of pounds of whiting while keeping his family and crew safe. The *Midnight Sun* is also one of the most photogenic boats in Gloucester Harbor.

CHAPTER 27

Magnuson Reauthorization

The effects of Magnuson reauthorization were not felt at the local level immediately, but everyone knew most management plans would have to be rewritten. Currently, fish quotas were indirectly managed by effort control. You could only fish so many days and keep so many of certain kinds of fish on those days. Some of the plans were draconian, but progress was being made. So, it was with great frustration that the council entered the period of 2008–2009 having to craft entirely new forms of management plans, the details of which had not been spelled out yet. The council did know that Limited Access Privilege Programs (LAPPs), government speak for individual fishing quotas, required an affirmative vote of two-thirds of permit holders and the chance of that occurring in New England was zero. NMFS was pushing sectors, having their legal counsel make tortured arguments that, because the quota was given to the sector and not the individual fishermen in the sector, that the sector system was not a LAPP. Numerous hours were wasted on philosophical analogies involving ducks (walks like a duck, talks like a duck) or in one fisherman's public comment, "If it smells like a skunk, has a white stripe down its back like a skunk, under NMFS's interpretation it's a racoon!" We were getting nowhere fast, and these new plans had to be in effect May 1, 2010. While this was going on, scientists continued to turn out single-species stock assessments, which now would be used to produce fixed-number quotas for the new management system. The old system, which used these numbers as targets and which was producing positive results, would be abolished.

Daniel began his graduate schoolwork after Labor Day and would no longer be available to crew on a regular basis. I was forced to hire people off the street. I employed an endless stream of characters. Zero tolerance for drugs and alcohol was considered a suggestion. Lest one think this was a function of only the younger generation, the most bizarre of the lot was an over-sixty-year-old. That man filled his thermos with rum with a little coffee for coloring. When I dumped the thermos out at sea one day, he threatened me with a knife at three in the afternoon. Apparently, delirium was setting in and without his daily fifth of rum, he needed a drink . . . bad. Later, he further distinguished himself by getting in a fight at high noon on the street on Hampton Beach with his daughter, who was trying to disarm him of a bag of grass. He was finally gone for good in March when he borrowed some guy's van and did donuts on the ex-wife's front lawn in Maine. Maine wisely remanded him to state prison. The last I heard from him was his daily phone call requesting his last check be sent to his psychiatrist's office.

The economy was humming and everyone, including the Yankee Fishermen's Coop, was having trouble getting help. The coop manager, Bob, hired thirty people at the beginning of a week and by the end of the week only one remained. Numerous men and women came from Eastern Europe on work visas to work on Hampton Beach. Some found their way over to the coop and were hired. They told their friends and others followed. We called them Russians because they had grown up under the Soviet system and were required to learn the language, but they came from many different countries. They were all conscientious and very hard workers. However, they did not understand that old animosities were not part of America's immigrant story. One day, I was talking to Bob, waiting for my fish slip when we both heard loud screaming from the warehouse. Two of the men were brandishing knives and fish picks screaming at each other. We received a rough English translation from a non-combatant that one man had accused the other of being related to people who impaled his great uncle Vlad in the 1500s. Bob had to explain that old country animosities are left in Europe. Finally, they put down the weapons, shook hands and went back to work.

Out of this group, one man had a green card or as he put it, a real genuine green card, not a genuine fake green card. He called this winning the lottery. People with work visas cannot work on boats, but those with green cards can. I hired the man and once I learned his limited knowledge of English was gleaned from John Wayne movies, we were able to communicate with English and sign language. He was an excellent worker but ran afoul of the law mainly because of not understanding the rules. In his country, if the police stopped you, you asked how much it cost to make the problem go away. Of course, here, when, say you were stopped for a traffic infraction and offered to pay off the officer, you were arrested for bribing the police. All these men drank vodka like it was water, but for the most part did not drive after imbibing. However, they called beer yellow water and did not consider it alcohol. Therefore, most had a DWI within six months. My crew was no exception and needed a daily ride.

While fishing off Boston in June, I saw the telltale red stripe of a Coast Guard boat approaching, but it did not look like anything I had ever seen. The boat was an icebreaker from Michigan on its summer cruise. The crew had just graduated Coast Guard boarding school and aimed to try out their newfound knowledge on the *Ellen Diane*. All the so-called Russians had an innate fear of men with guns and uniforms and seemed to possess some magic power to become invisible. We had been boarded before and no one ever asked the crewman anything. This time would be different, and after going through all my paperwork and safety equipment, the boarding office began asking the crew questions such as when the last time was you donned a survival suit. The crewman called survival suits tomato suits because they were red and had no idea what donned meant. I was trying to mouth the words the crewman knew from behind the boarding officer who was becoming very agitated that he was not being answered. Finally, the crewman answered I put on tomato suit last week. The boarding officer hissed at him, "Learn English, it will save your life!"

Next, the officer ordered me to haul up the net. The law says you can request, but not order, the net to be retrieved. I had a specific place in mind to haul back and refused. The boarding officer radioed the captain on the ice-breaker and stood down fuming. This was not going to end well, because I was pissed, and this guy clearly wanted to make admiral before he was twenty-five. About twenty minutes later, we hauled up the net so the boarding officer and his team could measure the codend. The legal size is six and a half inches, but I was using oversize twine of about seven and a quarter inches to let out as many cod as possible, as well as small flounders, which were very cheap at the moment. The measuring device is a wedge-shaped paddle attached to a twelve-pound cylindrical weight. Four enlisted men were holding the twine. I quietly warned them to move their feet. One asked why and I said you will see. The boarding officer stood with his feet directly under the weight and dropped the wedge in the twine. The wedge went right through, and the weight crushed his big toe. The officer turned red; the enlisted men tried to stifle their snickers. This happened several more times before the boarding officer hobbled back to his raft and departed with enlisted men in tow. I have a lot of respect for the Coast Guard and only had two problem boardings in my life. When we got into port, I phoned the executive director of the council, Paul Howard, to request information on how to file a complaint. I knew Paul would be upset. Paul asked me to give him a couple weeks to handle the issue. Sure enough, almost two weeks to the day, I received a phone call saying the problem had been handled. I asked how and did not receive a direct answer, only that he hoped the boarding officer enjoyed his tour of Nome, Alaska, in winter.

Just before a council meeting in February, I received a phone call from my son Eric wanting to know what the weather that night would be one hundred miles southeast of Boston. My son had worked his way up from engineer and was now alternate captain of one of Boston Towing and Transportation's big Azipod tugboats. Each vessel carries two captains who work alternate six-hour shifts. Eric was captain while the other captain slept. Currently, at my house the wind was thirty to forty miles per hour from the northwest and the temperature was ten degrees and dropping. Eric wanted to know sea conditions and if I thought the wind would drop overnight as the weather service predicted. I told him the wind would keep up through at least noon of the following day and seas were probably ten to twenty feet gradually dropping to ten feet with heavy icing conditions. I went upstairs and pulled my offshore chart from the back of my closet. What was going on? Eric's boat was large and new but was still a harbor tug. First, I asked if they had mallets to break ice. I knew Eric had little knowledge of the danger of ice, and his captain lived outside New England. Ice can build up on the vessel's superstructure, radically altering the center of gravity. Left unchecked, a boat can suddenly flip over and sink. Trying to appear nonchalant so as not to upset my wife, I asked why he was headed for Georges Bank in the middle of a gale. The response was that a liquified natural gas tanker had lost its engine and was drifting towards Georges Shoals. They were going to try and put a cable on the tanker before it drifted onto the shoals. He also wanted to know what the depths were, as they did not have offshore charts. I responded that the charts showed several areas of twenty-eight feet and more areas around thirty feet. Also, I said the shoals shifted constantly, and the chart was most likely not accurate.

Captain Eric Goethel (five foot, eleven) inspecting the propellor of the Boston Towing and Transportation tugboat *Justice* in the boatyard, 2009.

He replied it did not matter, as the tanker drew more than thirty feet. This conversation got worse with every sentence. Eric's tug, *Freedom*, would arrive around 10:00 p.m. and attempt to take the 930-foot *Catalunya Spirit* in tow. There were two goals here: keep the ship from drifting into the shoals and keep any gas from leaking out. There were two schools of thought about the catastrophic break up of an LNG. Eric's belief was the gas would burn off like a flare from an oil well. Others believed there would be an explosion of catastrophic proportions that would incinerate everything within three miles. The fact was, no one knew the answer because the event had never occurred. Eric did not seem concerned, as they were steaming with one other tug, and he assumed they would be back in port by the end of his shift on Tuesday. The boats would be out of cellphone range but in touch with the Coast Guard.

I had a council meeting the next day in Portsmouth and did not sleep all night. One of the members of the council is a Coast Guard liaison who comments on safety matters and rule enforceability but does not vote on council decisions. I sought out this officer and began to tell him the situation when he

Capt. Eric at the helm, joystick instead of a wheel and microphone instead of a hand-held VHF radio! A tug that can go sideways and spin in circles because of its Azipod unit (combination rudder and propeller), 2007.

suggested we go out in the hall. He knew what was happening, but senior officials were trying to keep it off the news. Government officials were afraid people on Cape Cod might panic. He could tell me that in a heroic feat of seamanship the *Freedom* had transferred first a line and then a tow cable to the stricken ship. The ship had no power whatsoever; the crew were using flashlights. He described how Dave, the captain, maneuvered the tug under the stern of this enormous ship in ten-to-twelve-foot seas, and guided by Eric with a handheld radio who was giving orders to the deck crew, used the laws of physics to their advantage. Usually, the heavy wire towline is hauled aboard by a capstan. In this case, the crew of the ship deployed steel cable from an emergency towing winch and through a complicated set of maneuvers, the *Freedom* was able to secure the towing wire. After two hours, the *Freedom* finally slowed the tanker's drift and gradually began to get the big ship turned into the wind. Two hours after that, they began to make very slight forward progress. If the tow cable did not break, the *Catalunya Spirit* would avoid the Georges Bank shoals.

This problem was caused by a computer-driven engine and exposed a big problem in modern machinery. The computer had detected a problem in the main propulsion unit and shut down the engine, but the only way the engine could be restarted was to restart and reprogram the computer as well as repair the original engine defect that had caused the computer shutdown. Once the engine cooled off, it required massive power to turn it over. Engine and computer experts were airlifted onto the ship but would be unable to start the engine for days. Older engines like mine had no computer. They were strictly mechanical, and if you had a battery, you could start the engine. I continued to make motions and debate the issues of the day, always keeping an eye on the Coast Guard liaison for a signal to chat. When Ellen arrived late that afternoon, she immediately engaged the liaison and they both left the room. Everyone assumed something was going on but still had no idea what. The story had finally leaked, and council members knew there was an LNG adrift, but that was all. By that night, the wind had abated, and the *Freedom* was towing the tanker at about two knots per hour into Massachusetts Bay, but a new problem loomed. A low-pressure system was approaching, and gale-force winds were expected to pick up from the southeast. This meant there was a possibility the ship could drift onto the

Disabled 933-foot *Catalunya Spirit* liquified natural gas tanker being towed by the tug *Freedom*. See the tow line attached to the stern of the tug. Georges Bank, Feb. 12, 2008.

Catalunya Spirit successfully under tow by the tug *Freedom*, thirty-three miles east of Chatham, MA, 2008.

shore of coastal Massachusetts. A decision was made to tow the ship back out to sea. This game of towing in and out based on wind direction continued until a large, offshore salvage tug, the *Atlantic Salvor* arrived from Philadelphia and the *Freedom* was relieved of its tow. On the third day of the council meeting, the Coast Guard liaison asked to address the meeting. He laid out the story of how my son was part of one of the greatest rescues in modern times, but that the crew would never receive recognition because the story was squashed to avoid public panic. The council and audience rose to give me a standing ovation. I was intensely embarrassed as I had not done anything. The captain and crew of the *Freedom* deserved the accolades, but they would never come. After five days, the engine was finally restarted, and the *Catalunya Spirit* quietly slipped into Boston, unloaded its cargo, and just as quietly sailed over the horizon.

After a year of strenuous work, the council finally had an incredibly complicated solution to comply with the revised Magnuson Act. They would impose a voluntary sector-management plan while having a parallel plan called the "common pool," based loosely on the current effort-control system. Since sectors were voluntary, the feeling was the program would be isolated from the LAPP requirements. To make sure everyone joined the "voluntary" sectors, the rules for the common pool, derisively referred to by fishermen as the "cesspool," had trip limits and day restrictions so onerous that only people with no quota, now euphemistically called "percent sector contribution," would put their permits here. This was the most complex management plan ever contemplated and would become known as Amendment 16.

At about this point, I received an email requesting my participation in an initiative called Baltic 2020. Basically, a group headquartered in Sweden, disillusioned with the negative consequences on Baltic Sea cod stocks, wished to fly me to Denmark to discuss the highly successful American days-at-sea program. I almost choked on my beer, as the council was chucking days-at-sea for a quota management plan. After consulting with the executive director to make sure there were no prohibitions on foreign consulting, I signed on. I was cautioned to make sure I expressed the fact that my views were my own, not those of the United States government. I found Copenhagen charming and the people I held discussions with knowledgeable and willing to put in the hard work necessary to radically change a management system. I did wonder how they were going to get compliance from more than ten different countries, when New England found it hard to get five states to agree. I was very well wined and dined and returned to Hampton figuring I would never hear another word from the group.

In April 2009, the council was informed NOAA administrator Jane Lubchenco would attend the April council meeting to discuss sectors. On the day of her arrival, agitated advance people swept through the meeting room scooping up all coffee containers made of Styrofoam or paper. Apparently, Jane was highly offended by any cup that was not reusable. Most people would not have cared if they brought in a big urn and ceramic cups, but they just swiped our coffee that we had paid the hotel usurious rates for and did not replace it. People were still grousing about the coffee when the NOAA entourage swarmed into the room. She got right down to business. She wanted sectors passed and wanted it done now. "Pass sectors or I will appoint people who will!" was her emphatic message. State directors, who are not appointed and represented a third of the council, just kind of rolled their eyes, but the rest of us got the message loud and clear. While we were technically appointed by the Secretary of Commerce, NOAA was the parent agency of the National Marine Fisheries Service within Commerce and held considerable sway over appointments. I thought she was impressed with her own importance and believed the president could have picked someone with more diplomatic skills to run the agency. At any rate, she would not have to wait long, the final vote on Amendment 16 was scheduled for the June council meeting.

CHAPTER 28

Treacherous Politics

Right after the council meeting in April, I received a second invitation from Baltic 2020 for Ellen and me to fly to Brussels, Belgium, headquarters of the European Union, for a meeting with most of the fisheries ministers of Europe in late May. Baltic 2020 had written their new management plan for the Baltic Sea cod fishery and wished to present it to the assembled ministers. Europe manages fisheries at a political level. They receive scientific advice, but ministers vote on the plans. Generally, they split along old religious lines with predominantly Catholic countries wanting higher quotas and predominantly Protestant countries more willing to accept scientific advice. I was informed that the two landlocked countries with no dog in the fight often cast the deciding votes after extorting agricultural favoritism for their countries. We were told to bring formal clothes as there would be a state dinner on the final night.

I was a little concerned, as I was a fisherman who was explaining an American plan the Americans were about to scrap. Ellen and I had free time and visited several historical sites in Belgium and ate dinner nightly in the old town section, which had amazing seafood restaurants with equally jaw-dropping prices. I had meetings daily that could best be described as meet and greets, sometimes with the European Union Fishery bureaucracy, and other times with groups of Baltic state ministers and with the heads of some Swedish fishery trade groups. The night of the dinner proved to be one of my more interesting experiences. The dinner was held in a large hall that was very ornate and formal. Men wore tuxedos and business suits and woman mostly had formal gowns. I would guess roughly twenty countries were represented. Most ministers spoke English, but a few brought translators. One minister attracted a lot of attention as he had a very beautiful translator about twenty-five years old surgically attached to him for the entire evening. I can guarantee you gossip is not only an American pastime! Formally attired waiters moved through the crowd with hors d'oeuvres and trays of wine. I soon learned that the European Union is a union that runs three glasses of wine deep. After that, the snippy comments began about other countries' history and lack of adherence to the fishing rules. Several people wanted to be my friend, but what some really wanted was to enlist me in an alliance against some other groups. I kept reminding everyone I was a fisherman with no ties to government and thus had no thoughts on political alliances. Dinner was served and food seemed to cool the political passions. Dinner was followed by a few speeches, most mercifully short, and thanks for my contribution. Baltic 2020 personnel informed me the ministers would vote in the coming weeks and I would be kept informed. The plan was never voted in, falling apart around the issue of enforcement, as I had warned people.

After seeing how potentially treacherous the political landscape was in Europe, I hoped the council could produce better results at their June meeting on Amendment 16. This meeting was being held in Portland, Maine, at the Holiday Inn. Council meetings are required to have large meeting halls capable of holding at least three hundred people. While the room was air-conditioned, the sheer volume of people soon overwhelmed the system. This was one of the most complicated amendments from a procedural point of view. Over fifty votes would be taken, with items voted on early often being impacted by later votes. Under the national environmental policy act (NEPA), most decisions required several analyzed options, thus the possible permutations of the outcome were too large to wrap your head around. Worse, under Robert's Rules you had to be on the prevailing side if you wished to make a motion to reconsider an earlier vote based on the outcome of a later vote. This meant being in the unenviable position of voting for something you opposed to be able to reconsider the vote later. Some council members had a poor grasp of Robert's Rules, and the audience had virtually no grasp at all. What could possibly go wrong? Further, the assistant administrator of NMFS flew in from Washington. Officially, he was there to observe; unofficially, he was there to intimidate wavering council members. Public comment would be very limited. This angered the fishermen, who mostly failed to understand that comments should have been made at public hearings, especially if one thought all the alternatives were unpalatable. An overheated room full of overheated people with more work scheduled than could possibly be accomplished in the allotted time; it was time for votorama to begin. The morning votes were relatively straight-forward, and the council progressed well, but the after-lunch votes were where the rubber would meet the road. Every council member was given a chance to speak once on a motion, and the audience allowed two speakers for and two against. Three votes would be extremely controversial: how the fish were allocated, how much monitoring of the fishery would be required, and who would pay for that monitoring. Monitoring came first. Some groups wanted 100 percent monitoring of boats and unloading, having long memories of the last failed quota management system in the late 1970s. A compromise was struck that would garner a 10–6 vote, but this sowed the seeds of discontent. Large vessels with large amounts of fish unloading in centralized ports relatively few times a year could probably survive, but small boats unloading in geographically dispersed regions daily would pay costs higher than the value of their fish. Worse, if the fishery was a limited access privilege program (LAPP), cost recovery was limited to a small percentage of your gross, but these were not LAPPS by an earlier vote, so cost recovery was the total amount of the cost. A motion to reconsider was made. Pandemonium ensued as the vote had been a hand vote and the maker of the motion could not prove he was on the prevailing side. Finally, the chair called for a break while council leadership and NMFS legal counsel wrestled with how to proceed. After the break, the motion to reconsider was ruled valid but failed on a roll call vote 9–8 with the chair voting to break the tie.

Next, we moved on to who paid for this. Here is where I made a strategic mistake. NMFS leadership announced that for the foreseeable future, NMFS had the money to pay for the monitoring, but the council should vote that fishermen had to pay. NMFS is not granted taxing authority, and that fact should have been challenged at that point, but there was too much going on and I let it pass. I lost again on a 10–6 vote. Now it was obvious that representatives of the large boats, the ENGOs and recreational fishermen had decided this was going to pass, but why? The answer would be revealed shortly. I was sitting next to the Massachusetts state representative. He had a long list of how he was to vote but had joined the losing side on the last several votes over the fairness issue. His state director came out from the audience and took the man's seat to vote on behalf of the state on the allocation issue. This had never happened before either. I made a motion for the option that put everyone on the same timeline for allocation 1996–2006. The motion was seconded, but the first council member to speak made a motion to substitute: the Cape Cod Hook Association, a group of fishermen on Cape Cod heavily funded by ENGOs. The association had fished with lines of baited hooks, which had been an experimental sector under Amendment 13, 1991–1996, the recreational fishery 2001–2006, and everyone else 1996–2006. This was grossly prejudicial; the hook association would control 20 percent of Georges Bank cod, an amount they had never caught outside that period, and the recreational fishery would get over 30 percent of Gulf of Maine cod even though their historic share had been 10–13 percent because the rest of the fishery had been so constrained during the allotted period. The room erupted with angry shouts, and I thought a riot might ensue, but we soldiered on. The play now was obvious: Massachusetts wanted to protect the recreational fishery and the hook association was politically powerful in state politics. Several large-boat players wanted sectors, figuring they could buy up enough of the fishery to survive, and the EDF representative on the council would vote for anything that put sectors in play. I called for a roll call, but I already knew the outcome: I would be on the losing end of a 10–6 vote. Fishing permits were now a commodity to be bought and sold by whoever had the most money. You would now survive based on how many permits you could buy to control enough quota to continue fishing.

I had feared this scenario and had requested the coalition's Magnuson expert brief me on how to file a dissenting opinion with the secretary of commerce. I had only witnessed one person do this since the inception of the Magnuson Act. Basically, there is a three-part process: you must announce before the final vote that you will be filing a dissenting opinion, then you vote against submission in a roll call vote, and finally write your letter citing specific parts of Magnuson within fourteen days. My friend, who had been on the New Zealand trip, planned to do this as well but did not get the steps down correctly. After the vote, I was informed by the executive director that I could not use council staff to research my letter.

My letter was outlined in my hotel room that night before the meeting was finished. I was so certain the allocation vote had violated several national standards that I stuck mostly to that subject alone, figuring it would force the secretary to remand the amendment to the council. The coalition's representative from Washington provided me with all the appropriate citations from

Magnuson and the letter went off to the secretary of commerce, the council, and the NH congressional delegation. Then I waited and waited . . . and waited.

My crewman Jeff and I resumed fishing when the area off New Hampshire opened in July. I had been contacted by New Hampshire Public Television, which ran a show named *Windows to the Wild*. Their narrator and cameraman wished to film a show on fishing for cod. They shot some beautiful sunrise scenes and filmed us setting and hauling the net, as well as an observer measuring the catch. This, along with narrative explanations was edited into a half hour show. The episode proved to be one of their most popular and was shown many times. My wife still shows the episode in her oceanarium, as it is one of the most concise and accurate explanations of how dragging works. Other fishermen good-naturedly ribbed us about being movie stars.

Late in the fall of 2009, Jeff asked for a week off and Dan made a guest appearance. An event occurred which was distinctly less cinematic. Dan and I were using my rockhopper net to catch our eight-hundred-pound daily limit of cod about twelve miles southeast of Hampton. We had left early because Dan had to give a lecture at two o'clock at SMAST in New Bedford, which is about a three-hour drive from Hampton. We had been catching our cod in two tows and I figured if we were in by 10:00 a.m., Dan would be all set. Our first tow produced about three hundred pounds. On the second tow, the boat slowed down coming off the bottom and I figured we had picked up cutoff lobster traps. I tried to go over the bottom again hoping the traps would hang in the rocks and break off the net, but no luck. As we hauled the net up, the boat was pulled backwards towards the net and the towing wires were narrowly spread, never a good sign. When the doors were about halfway to the surface, the wires began to spread and rise on their own behind the boat. What was going on? Suddenly Daniel pointed behind us in awe as a very dead whale blew out of the water. Everything dies and whales are no exception, but what is different is how long these giant animals take to decompose. This was my third encounter with a dead whale, and they were all difficult and dangerous. The whale was in the net headfirst, or more accurately, what used to be headfirst. It was a white gob of goo about thirty feet long with a nauseatingly disgusting stench. My last encounter, I had tied a rope to the tail and let the water pressure pull the net off the whale before cutting the beast loose, but this had no tail; it had rotted off. The remains had not made it into the end of the net, so we worked the codend around, cut it off, and lifted the fish into the boat. Next, we strapped the beast across the stern, and Dan held me upside down by my feet so we could split the net with a knife. There were bits of white goo falling off the net on the reel onto us and the ocean was a mass of bits of goo and whale oil. Even the seagulls left! We were both gagging, and Dan was becoming concerned about the time, but we could not move until the whale was gone. After what seemed hours of being suspended with my nose inches from this awful mess, I determined the net had been cut apart enough to let the thing go free. I went up to put the boat in gear and told Dan to motion me when the carcass started to move. The whale moved and Daniel motioned for more power. What neither of us knew was that the remains of the pectoral fin were stuck through a hole in the bottom of the net. I kept advancing the throttle and finally the boat lunged forward. The whale rolled with the pectoral

fin stuck up in the air like an extended middle finger with most of the bottom of the net attached. Daniel would not have time to shower and shed his fishing clothes and rushed out of the parking lot when we got to the harbor. He still had bits of whale goo in his hair. Dan arrived just in time for his lecture and explained to those assembled what had occurred. He told me it was the quickest lecture ever, with no questions as everyone rushed for the exit stifling coughs and gasps from the smell. Dan arrived back in Hampton just as I finished

"Oh @##*, how are we going to get this out?" Retching from the smell of a decaying carcass of a whale. Photo by Dan Goethel.

replacing the missing twine and reattaching the codend. We would end up taking our clothes and raingear to the landfill, as no soap on earth could counter the smell of a very dead whale.

In the fall of 2009, I was invited to testify in front of the House subcommittee on fisheries and oceans about catch shares. At least somebody had read my dissenting opinion. I would have five minutes but could submit longer written testimony. My message was simple: we had a system that was working, and we were replacing it with a system based on greed that was supposed to somehow magically provide sustainability. The program would eradicate small-boat communities like those in New Hampshire, be frightfully expensive to administer, and decimate recovering fish populations, because a handful of large boats (generally defined as over seventy-five feet) staying at sea for multiday trips and taking entire schools of fish was a very different beast from boats spread out up-and-down the coast taking only small amounts of fish and returning to port on a daily basis. My calculation was that both fish and fishermen would be destroyed in five years. I was wrong: Gulf of Maine cod were destroyed in less than three years. If you have never testified in front of Congress, I can assure you it is a humbling experience. The committee sits around a horseshoe-shaped desk which is raised above the surrounding room. This is so press photographers can kneel in the center of the horseshoe and take photos without being seen. I was part of a panel of five speakers who sat at a table facing the horseshoe. Behind the panel was the gallery where interested citizens could sit and observe the proceedings. Each of the panelists had a microphone with a traffic light system at the base. When the red light was on you could not speak, yellow light meant thirty seconds, and green was go. Luckily, I was not the first speaker. I had been warned that people freeze when the light turns green. They were right; I felt like someone cut out my larynx. I swallowed hard and stalled fifteen seconds before I finally started croaking out my talk. The whir of the cameras is a tremendous distraction, but I did my best to ignore them and make eye contact with the chair. Clearly my message caught several committee members by surprise, and I got a few questions. Those answered, I was done, but not for long; I would be back six months later. April 2010, there was another round on the same topic. This time, there was a fisherman from Maine on the panel next to me. I warned him about the gag reflex. He was a very jovial man and said he was sure he could handle it. When his light came on, he froze for thirty seconds. We laughed about it afterwards; there was just something about the setting that sucked the life right out of your chest.

As spring rolled around, I was still awaiting a response to my dissenting opinion. I contacted Senator Gregg's staff to see if they could track down what was taking so long. The answer came back that I would receive a written response soon. This I did get two hours after publication of the final rule announcing Amendment 16 was already approved. I had put my name forward for a third term on the council in March. Now I wondered why I had bothered. I assumed I would not be reappointed, as I had clearly tweaked some very powerful people. I was stunned when I received a call from the regional administrator in June telling me I had been reappointed. One of her statements was "You must have some very influential friends." I guess so, but I can only surmise who stood their ground on my behalf in the upper echelons of NMFS.

The bureaucratic nuts and bolts of sectors proved every bit annoying as expected. Now you had to register to fish two days in advance of departure so you could be assigned an observer or monitor. On the way home, you had to phone a dockside monitoring company so a monitor could watch you unload. If the monitor was not present, you had to wait for their arrival, letting your

David and Daniel picking a load of whiting.

fish bake in the summer sun. The coop could not begin weighing your fish until the monitor was watching. Numerous disagreements arose when the warehouse workers had different tallies than the monitors. This was big brother on steroids, and all it proved was how many people could be employed to foul up a fishery. I could not wait to go whiting fishing and get away from this circus. And so things stood in mid-fall 2010 when I went towing for whiting and herring with my current crewman, Jeff. Jeff had graduated UNH with a degree in engineering and a large college loan. He went to work for me while he tried to find a job. He was a good, conscientious crewman who overcame chronic seasickness. September 15 was just like so many fall days. While I was towing, I had been reading the sea herring stock assessment for the upcoming council meeting. This would be the last item in my long-term memory after I fell off the pier that afternoon.

CHAPTER 29

Being Put Back Together

Remembering all the previous stories and events kept me occupied almost the entire length of my stay at Beth Israel. Initially, stories came in jumbled waves, something from my childhood followed by a fishing story perhaps followed by events in fishery management, but over time I became capable of sorting the memories and filing them by period. My brain was clearly functioning more normally, and the neurologist was pleased to announce on the fifth day that all the blood had been reabsorbed. He could not guarantee a future, unforeseen tragic event, but right then things looked good. He also waxed somewhat philosophical, stating that very few people left this ward in anything but a wheelchair or a body bag. Science could not predict why some people fall two feet off a stepstool and die and others fall twenty feet and live. I would be part of the 2 percent who were success stories. "There must be something for you still to do on earth. Keep that in mind." He would check with all the other teams, and if no one objected, I would be transferred to a rehab hospital the next day.

Around noon time on September 21, I was placed on a gurney, rolled to the elevator, brought to the ground floor, and discharged to an ambulance for the ride to the Whittier Rehab Center. The ambulance had seen better days and the ride was frightful. The streets of Boston are a series of potholes interspersed with manhole covers driven over by people who should be committed to an asylum. There are only two pieces of equipment required on vehicles: a horn and a window crank (so you can roll down the window and extend your middle finger). The driver also needs a set of leather lungs to scream obscenities at whomever he just cut off. I could not see the driver, but I could hear the horn and the voice.

Arriving at the rehab center, I was transferred to a room and placed in a bed. The ride had sapped my energy and I dozed off. I was roused by a collection of staff who introduced themselves. Doctors would continue to monitor my physical progress; mental health people, both psychiatrists and psychologists, would monitor my mental progress; and an assortment of physical therapists would begin my rehab. They expected my stay to be thirty-five days. My response was "I have a meeting in seven days. I need to be functioning by then." These people were there to offer encouragement, so no one said that was impossible, but that was the look of their collective faces. Probably to temper my expectations, I was placed in a wheelchair and given a tour of the parts of the facility I would have to master. First up was the kitchen, where I was to cook breakfast, pour myself drinks, and remove dishes from the cupboard above the counter. Next up a room with a car where I would have to get in and out of the passenger seat. Finally, a bathroom where I would have to get on and off a toilet and into and out of a shower. We would work on all three every day. I had my work cut out for me.

The kitchen was the easiest. I only had my right hand, because my left hand was suspended in a sling and taped firmly to my body. I could barely move my left fingers but not grasp a pen. Moving said fingers produced great pain in my broken scapula. Think about pouring a glass of tonic from a two-liter bottle. You must somehow pin the bottle down so you can unscrew the cap. I pinned the bottle with my right hand and used the tips of my left finger to turn the cap. Next, cook breakfast. Getting the eggs out of the fridge was easy; cracking them, not so much. I guessed I would learn to love scrambled eggs complete with some shell. Buttering toast was another challenge, but by the end of day one I could handle the kitchen. Getting in the car was a real challenge, specifically transferring from the wheelchair to the car seat. I was given some pointers and surrounded by staff. Good thing, because otherwise I would have been in a heap on the floor. Getting out was easier, as I could use my right arm to brace myself on the door frame, plant my right leg on the ground and pivot to the chair. The bathroom, however, was a nightmare, especially the shower. I had to lift myself over the rim of the bathtub. Putting my broken left leg in first left me unable to get the rest of me the shower. Using my right leg first required lifting my left leg higher than it would go. It took four days to come up with a routine. I was glad they never turned on the water; I would have drowned! By now, other than my left foot, which kept growing for no apparent reason, swelling everywhere else was slowly subsiding. My left side from shoulder to groin was a hideous shade of black, purple, blue, brown, and yellow, but, miraculously, I had no broken ribs or ruptured internal organs. My left face and ear looked like I had been mauled by a pit bull, and the left side of my mouth hung slack, which made speaking coherently difficult, though my teeth seemed to be migrating back into position on their own. I needed drops in my eye every twenty minutes and was informed that once I was strong enough, I would need a gold weight placed in my eyelid and would need to have my face surgically "hung up" in my eye socket. This operation was tentatively set for December.

The mental health people seemed strange to me. Not strange as in unpleasant, but they asked the oddest questions. One wanted me to name my ex-girlfriends. Another asked me to describe my job. Another time, I was asked to describe the last item I remembered reading. That question produced a lengthy dissertation on the herring stock assessment I had read my last day fishing. The elicited response was that the man had no idea what I was talking about, but clearly my comprehension and long-term memory were outstanding. Great, now if I could just remember the man's name. which he had provided fifteen minutes ago! The upshot of all this mental poking and probing was that the long-term memory was normal, i.e., unchanged, short-term memory for names and numbers was poor, and I had moderate control of anger management, which might be permanent or a function of blood-sugar levels which had spiked from the fall. Not bad all in all.

On the day before the council meeting, I called Ellen to come pick me up for the ride to the meeting. I told the staff I wished to be discharged. Ellen told me she needed to contact the hospital to order a wheelchair and would call back. The staff told Ellen that having a wheelchair delivered would take a few hours, and she should arrive around 3:00 p.m. Ellen arrived to find me in the lobby filling out forms. Using my right hand, legibility was not a strong suit. Among other items, I had to certify I was moving to a handicap-accessible facility. Which I was—the council had reserved a handicap-accessible room at the Hotel Viking in Newport, RI. I was also told to get out of the car every hour to improve blood circulation. An orderly loaded a large, flat box into the back of the car. We both assumed it was the wheelchair folded up and never opened the box. Bad assumption. The staff tried one last time to convince me to stay, but I was determined. My steely determination and fierce willpower had got me this far; I could handle three days of meetings. I wheeled myself out to the car, and with what I had learned, loaded myself in the passenger seat.

The ride was not as miserable as I expected. We stopped twice and I got out and leaned against the car to stretch. I could not walk, and besides, the thought of using a roadside restroom gave my bladder fortitude I did not know I possessed. We arrived around 6:00 p.m. at the hotel, and a valet moved to greet us. Ellen explained the situation and the valet removed and opened the box from the back of the car. I heard Ellen exclaim, "You have got to be kidding me!" The box contained the wheelchair in pieces with an instruction booked stamped "Some assembly required." By now, the entire valet corps had assembled, and people were on their hands and knees putting together the wheelchair. I think it took about twenty minutes, but, at the height of the check-in period, we had tangled traffic for several blocks in both directions. Ellen was deeply embarrassed, but everyone was very pleasant about the issue. I was wheeled to our room, and we ordered room service. I had subsisted on hospital food since regaining my eating privileges; having real food was an unmitigated pleasure. Also, sleeping in the dark without all the hospital machines whirring in the background yielded me some actual sleep.

The next morning, Ellen ordered breakfast, assisted me with showering, and prepared to wheel me to the meeting room. The council staff had kindly seated me near an exit which was the closest to the handicapped restroom. Daniel's girlfriend, Annie, was part of the council staff and had shared the story of the extent of my injuries with people. I am sure everyone thought I would not show up. Scallops was on the first day's agenda. I just wanted to vote against consolidation in this fishery. Scallops had different rules than groundfish and a number of permits could not be placed on one boat. If consolidation was allowed, several permits could be placed on one boat and the other boats without permits could be sold or scrapped. The problem was that each boat had a ten-man crew which would no longer be needed. New Bedford, Massachusetts, has a large scallop fleet and these very lucrative crew jobs were vital to numerous families. Scallop-boat owners were doing very well financially; this was strictly about the rich wanting to get richer.

If the day proved too strenuous, I would leave that night. Several hotel staff held the double doors to the meeting room open so Ellen could wheel me in. A few worried scallop crewmen and captains filled most of the seats, but over in the far corner, several pro-consolidation lobbyists were holding court. One man was absent-mindedly flicking a pencil back and forth between his fingers. The murmur from the crowd caused him to look up. He snapped the pencil in his fingers. Ellen wheeled me to the table. I was at the very end. The man beside me on my right was the ASMFC liaison, Vince O'Shea. Vince was retired Coast Guard and was about six feet, six inches tall. Having him next to me reassured me; if worse came to worse, he could carry me out of the room. Several people surrounded me and offered any help necessary. I said that I might ask Vince to say a few words for me periodically because my voice was hard to understand, but that I was fine. Meanwhile, a lobbyist told Ellen that I looked very sick and probably should go home. Ellen smiled sweetly saying being here was all I had thought about since I fell, and I was not going anywhere. The pro-consolidation group looked like they had seen a ghost.

As the climatic vote grew nearer, I scrawled four short sentences on a pad. First, I was literally and figuratively glad to be here. Second, I thought a vote to allow consolidation violated the national standards. Third, there were no biological advantages to consolidation, and fourth, I wanted a roll call. Ordinarily, I would offer a paragraph-length rationale for the two middle statements, but I figured all I needed in my condition was the essentials. Perhaps buoyed by my presence, several fishermen came to the microphone to tell how they had been ordered not to attend public hearings, and those who spoke out would be the first fired if consolidation passed. The whole ugly ball of wax began to unravel. Going in, I had handicapped the vote as a 9–7 loss, but some council members were clearly uneasy with what they were hearing. The roll call began, and I tried to make pencil marks as votes were called, but I could not see my own marks clearly enough. I had to wait for the official toll. The executive director announced the totals: seven in favor of the motion, nine opposed, chair abstains. I slumped in my chair as this had taken more out of me than I expected, but I was extremely proud of my vote and still consider it the most important vote I cast in my time on the council. Hundreds of fishermen from economically distressed ports, from New Bedford to Virginia, still have high-paying, head of household jobs as of this writing because of this one vote.

The rest of the meeting was anticlimactic, and by the middle of the third day, I was feeling worse. Sitting in a wheelchair all day is incredibly painful. Until now, I had never understood the pain that permanently handicapped people feel daily. At later meetings, numerous people would ask me why I put myself through that meeting. My response has always been you have to stand for something. I have always viewed life as black-and-white, good or bad, whatever you want to call it. Unfortunately, most of the world sees life through the prism of legal and illegal. Many people learn the hard way that legal and illegal are very different from what's just in some cases.

One last indignity awaited me. I had not considered how I would get into my own house. Ellen could pull the wheelchair up the two steps, but the door was not wide enough to accommodate the width of the chair. I had to crawl to get inside. Ellen brought in the chair and wheeled me to our one downstairs bed. I regressed for about a week, and the visiting nurse considered sending me back to the hospital. The main problem was my foot below my broken ankle. It was now about four shoe sizes bigger than my right foot and turning purple. An MRI was arranged where doctors discovered fourteen broken foot bones. They had started to heal but would never work correctly unless rebroken and set. I declined, I would rather limp than have more pain. Another problem we encountered was the lack of a shower on the first floor. Necessity is the mother of invention, and Ellen had me sit in a shower chair with my back to the kitchen sink. She used the dish sprayer to hose me down and then mopped the water off the kitchen floor.

Gradually, people began to visit. Most were concerned with interfering with our lives, but I was incredibly lonely. The fishing community had been working to collect funds to help us survive and to take care of my boat, which remained in the water. I was and still am extremely grateful for the numerous acts of kindness that helped us get through this incredibly difficult period. My friend Erik came by weekly, while my son Eric moved the *Ellen Diane* to Gloucester to have a hydraulic winch repaired. While in Gloucester, my friend Tom checked the boat daily, and other fishermen rotated through making sure the batteries did not go dead. Literally hundreds of people performed acts of kindness too numerous to mention. I was scheduled for plastic surgery just after Christmas and would start physical therapy six weeks after that. Doctors were telling me functionality was the goal, and I should apply for Social Security disability. My response was always the same: "America was built by people who work for a living; I am going to fish again!"

CHAPTER 30

Return to Fishing

By Thanksgiving, my scapula had healed sufficiently to free up my left arm. I kept the arm in a sling for comfort, but I regained use of my left hand for eating and writing. Several iterations of devices had been placed on my left ankle, but that, too, was improving. I could place oversized slippers on my feet and had shed the wheelchair for a walker. I could not go any distance because of weakness in both legs, but I could get from the bedroom to the bathroom or living room. Ellen had driven me to numerous doctor appointments in and around Boston, and, in general, people were pleased with my progress. I also never missed a council or committee meeting. Getting to and from meetings and in and about hotels was exhausting, but doable. I was scheduled for plastic surgery just after Christmas for two procedures. A gold weight would be implanted in my left eyelid, much a like a sash weight in a window frame. I still remember having various size gold weights taped to my eyelid to get just the right amount of weight to allow the lid to close without drooping excessively. The surgeon settled upon 1.2 grams of pure gold. From then on, I joked that I kept my savings right where I could keep an eye on it; eleven hundred dollars' worth of pure gold still resides there. The second part of the procedure would be more complex, but the plastic surgeon played down the severity. Basically, the attachment point for my left facial muscle would be severed, a piece of muscle removed, and the remainder surgically reattached in my left eye socket. This would pull up my lips and stretch the muscle, so I would no longer appear to have had a stroke. This was all supposed to take about three hours, and Ellen would drive me home. The operation went smoothly, but I had an adverse reaction to the anesthetic. My blood pressure dropped to a dangerously low level and stayed there. I could remember seeing people but being unable to move or communicate, like I was suspended in a dream. Nurses told Ellen that she should go eat lunch as I would not be released for a while. Ellen knew something was wrong and demanded to see me. As I have said before, my wife can be a force of nature if necessary and was soon seated beside me. She watched all afternoon as various compounds were injected to raise my blood pressure and get me out of the twilight zone. Finally, by dinnertime, I was discharged, and we were headed back to Hampton. I remember the horns and the lights of the cars, but as the saying goes, the lights were on, but nobody was home.

After about twenty days of post-operative recovery, I was placed in physical therapy to begin strength training. The goals were limited shoulder mobility and being able to walk, albeit probably with a cane. I went twice a week for two hours. This was grueling and painful work, but if I was going to fish again, it had to be done. Also, at the January council meeting, the council approved a measure that was a direct result of the cooperative research I had participated in on cod. That research showed that cod returned to the same area year after year south of the Isles of Shoals to spawn. Research by others had shown that both cod and haddock need complete darkness and quiet to successfully spawn. In fact, work in Scotland had shown that the darkness must be so complete for haddock, that red emergency exit lights had to be extinguished to induce spawning! A few of us had wanted a prohibition on night fishing for decades, and now we finally had the proof. Fishermen had two caveats to endorsing the three-month cod spawning protection area: first, that everyone—including recreational fishermen—be excluded, and second, that we be allowed to transit across the area because it was directly off Hampton Harbor. There had never been any limits placed on the recreational fishery, and while some backed the proposal, a number put forward tortured arguments as to why the recreational fishery should get a free pass. Commercial fishermen countered saying they would never back this proposal if the government had tried to ram the idea down their throat, but because they had helped to do the research, they backed it. A way to transit the area was worked out, and the plan was voted through. For once, both the fish and the fishermen had a win.

My son Eric brought my boat back from Gloucester in mid-March with the winch finally repaired. I asked Ellen to drive me to the local marina so I could see the *Ellen Diane*. When she saw how profoundly the sight of the boat affected me, I got an immediate lecture about not even thinking about trying to get aboard. I still had to use the walker to walk any distance, but the physical therapist had begun teaching me how to go up and down stairs and to walk with a cane. Stairs terrified me, especially going down, which was much harder than going up, but after months of practice, the methods became hard-wired in my brain. I still use what I learned today.

On a fine spring day in April which was more like summer, I called my crewman, Jeff, who was working at the coop, for a ride to the boat. Ellen was off doing one of her school presentations, and I wanted to try out my leg. I knew Ellen would not let me attempt to board the *Ellen Diane*, but I had to know if I was capable. I told Jeff I wanted him to break my fall if I started to slide down the ramp. I could tell he also thought this was a bad idea, and we must have looked ridiculous with him in front of me as I inched down the ramp, but ultimately, we were triumphant. I used the cane to limp to the rail of the boat and sat down victorious. I had thought about this for days and next worked on getting my bad leg over the rail using my good arm. Finally, I stood up on the deck of my boat. I was back. I had Jeff take a picture so I could show people. However, when I showed Ellen the picture, she was less than amused, and the physical therapist was not impressed, either. I did not care. I doubled down on the strength conditioning and set June 1, when Ipswich Bay reopened, as the target for my first day back. By then I would be back on the mooring, because we could only stay in the marina until May 15. Getting in and out of the skiff would be a real challenge, and I set to work meticulously planning how I would pull this off. I told one of the council members at the April meeting that I was planning to fish. This man worked with fishermen, training them how to fish more safely. A few weeks later, I received a box in the mail. Inside, were an inflatable lifejacket to wear while fishing and a safety vest to wear in the skiff with a note saying simply, "Fish safe." In mid-May, Jeff called me to inform me he had found a job in engineering and would not be back. I was sad, as I was counting on having experienced crew, but thanked him profusely for all his help and wished him well.

I placed a notice on the board at the coop looking for experienced crew and soon got a call from a past crewman who was working on a trip boat and wished to return to day fishing. The man had issues with the IRS and usually moved on when they found him, but he did know how to run a boat and was a hard worker. Also, my son Daniel had been discussing the situation with his PhD advisor, Dr. Cadrin, and was given a leave of absence for the summer.

Officially, Daniel told his mother he was coming home to assess if dad could do the job. I registered to fish on June 1 and received a waiver from carrying an observer. Good, one less problem. The crewman and I would go out in the skiff to get the *Ellen Diane* off the mooring. That way, if something happened the crewman could run the boat. We would pick Dan up at the pier. I would spend almost all day in my nice, overstuffed captain's chair. The crewman would run the winch controls and Dan would run the net reel. All I would do is drive. June 1 was a beautiful day, and the plan worked. We caught almost three thousand pounds of flounder. On June 2, I was assigned an observer. When I called the company to confirm the trip, I explained my situation and that I needed an experienced observer who would go with the flow. Instead, I got someone who was intimately related to the rear end of a horse. This guy liked to throw his weight around. But I am the captain of the vessel, and he would do as I say. We had what you call a situation. He demanded I climb up on the roof while he examined the raft. I refused for obvious reasons. Then he said he would terminate the trip. I responded with "Get your supervisor on the phone." "My supervisor is asleep." My response was, "he won't be for long." Next, I said I would call my wife and wake her up to get the number and she will call your supervisor on the phone along with every NMFS employee in our rolodex. By now, he was sensing he had overplayed his hand. Finally, he agreed to have the crewman accompany him. We left and he sat on deck all day pouting. I filed a complaint when I got in, the supervisor apologized, and I never saw the man again. Oh, and like it matters, but the fishing was quite good.

The three of us settled into a routine, and fishing was going better than I could have hoped for. The crewman wanted a couple days off for the Fourth of July, so Daniel and I decided to make a trip on the third. The whaleback spawning closure had ended on July 1, and we decided to make a short tow to see if the cod were still around. Naturally, I had an observer, this time a young woman from Rhode Island. I would run the winches and the boat; Dan would run the reel. We set out and started to mark fish on the bottom. After about twenty-five minutes, I had a definite cod mark and woke Dan up to put on his rain gear and haul back. In the time it took Dan to get ready, I had two more marks. We hauled the net up and the codend blew out of the water thirty feet before the doors reached the surface. Nuts, I wanted some cod, but this was ridiculous. When we had the net back to the boat, I guessed we had eight to ten thousand pounds of cod. Under the old management regime, I could have let some swim away alive, but with this plan you had to keep everything. We started putting the fish in the boat two thousand pounds at a time and throwing them up against the back of the wheelhouse to move weight forward. When we were done, the *Ellen Diane* was full of cod up to the top of the rails from the back of the wheelhouse to the stern. The girl asked if we were going home. Yes, after we clean all these fish. Do you want to help? No, I am going down below to sleep. Now remember, do not throw any overboard! For just a minute, I thought my head might explode. Next, I asked Dan how we were going to do it. Dan had a plan. He would fill three fish boxes, and I would cut standing up from the top box. I knew I could cut and gut, but pulling the gills would be an issue. I also did not know how long I could stand. We started from the stern and worked forward. Dan is big and very strong, but even he had a hard time pulling the gills on these fish. Some weighed over eighty pounds. I showed him how to cut the gills out of the biggest ones. Dan tried to load my boxes with ten-to-twenty-pound fish so I could pull the gills with my right hand. Unfortunately, most of the fish were over twenty pounds. We had started at 6.30 in the morning, and since this was July by nine it was hot. The observer came up on deck around 9:30 to inquire again when we were going in. My response: "When these fish are all cleaned and washed. If you want to help, we will get done quicker." "Nope, I will go back to sleep, but it is getting hot down there." Daniel just looked at me and laughed at the absurdity. By 10:30, we were done cleaning the fish. We had enough fish boxes to box and ice about half the fish. We would throw the rest under the net reel to keep them in the shade. The ride home would take about an hour and twenty minutes. Dan would have the fish boxed by the time we got in. I was in serious pain. This was not in the game plan. I found a bottle of water and a bottle of Tylenol. Four pills and a half hour later, I decided I might live.

I called Bob at the coop to tell him to get ready. Trying to be jovial, he said what did you do that for? I responded, "Don't ask," and he realized there was no humor left in my voice. Just then the observer came back up on deck and asked, "Mind if I ask you some gear questions?" For just a moment, I considered what would happen if an observer tragically disappeared at sea. Then I answered "Sure, we could not possibly have any form go unfiled." After an hour and a half at the coop, all the fish were unloaded, roughly eighty-five boxes. I picked up the slip the next day, 8780 pounds, my biggest groundfish trip ever. I kept the boat tied up for the rest of the week, both for me to recover and because we had caught 20 percent of our cod quota for the year in thirty-five minutes. Surprisingly, we were well paid for the fish, averaging about $1.80 per pound, as very few boats went out around the Fourth. Dan made $3,000 for seven hours' work. The other crewman was kicking himself for a while about taking time off.

CHAPTER 31

Epic Frustration

The fishing continued to be extraordinarily good and consistent. Just like magic, the cod left right before the opening of the silver hake (whiting) season on July 15. These days were short timewise, but intense. Typically, the first tow would be half an hour in duration, both to check the quantity of fish and so the crew could get a head start on boxing and sorting. The second tow would typically be an hour, and, if necessary, a third tow would last an hour and a half. The reason for lengthening the time out is that whiting is thickest right at sunrise and they spread out across the fathom curves as the sun gets higher in the sky. So, the first tow might have two thousand pounds for a half hour, the next tow twenty-five hundred for an hour, and the last tow twenty-five hundred for an hour and a half. Keep in mind the tows also contained red hake for lobster bait,

Still recuperating the summer after my fall, but I still have what it takes. Holding myself up with the towing wire, the crew unloading seven thousand pounds of herring! Now, how do I get back on the boat? Yankee Coop, Seabrook, NH, late summer 2011.

Capt. Dave relaxing between tows and finishing paperwork for the NMFS. Gulf of Maine, 2012.

some sea herring, and, in the last tow, some butterfish and squid. Every species has a trip limit, so you had to be mindful of not exceeding the limit with the concomitant waste of fish. Despite how fishermen are painted by many ENGOs, everyone I know tries very hard not to waste any fish. The seagull population is doing just fine and does not need a free meal.

By mid-August, the whiting begin to thin out and the sea herring begin to arrive. Herring spawn in late September and early October, but the time varies from year to year as does the place. We are severely restricted to where we can use small mesh mobile gear. If the herring come to us before the spawning closure, we have some big trips; if not. so be it. Herring is considered a pelagic fish, that is, they swim around off the bottom in the water column. However, right at sunrise, as spawning season nears, they drop to the bottom for several hours. Herring is very densely distributed, and one can often catch five to ten thousand pounds in a fifteen-minute tow. Fishing for herring is an art: set out too early, and you tow under the fish; tow too long, and you get more than the boat will hold; tow too short, and the fish are off the bottom for the day by the time you reset the gear. On a flat calm day, the *Ellen Diane*, loaded just right, holds a little over 120 boxes of herring. Numerous days this fall, we would catch close to our limit. Since you must box the herring and move the weight forward, you cannot put all the fish in the boat at once. Usually, we would split the bag twice, putting about four thousand pounds in the boat and then load some more fish, but you had to be careful. Herring float because their swim bladder expands as they are brought to the surface, but after a while, the fish expel the gas and then they become dead weight. There have been several purse seiners, which catch much larger quantities of fish, that have been rolled over by the fish sinking in the set. To guard against this, we would hang the next bag of fish by the block. That way we could load it in the boat, and if the remaining fish became dangerously heavy, leave the end of the net open and let the fish go. My crew was like a well-oiled machine. They could shovel and move a lot of fish in a short period of time. The crew liked herring fishing because they did well financially, and because the days were short. We were usually unloaded and back on the mooring by noon. One fine day, Ellen came to the coop to take pictures of the crew shoveling fish

Dan always did love to sit right in the middle of the fish. He was perpetually covered with scales, 2011.

Dan Goethel finishing up boxing twelve thousand pounds of herring, 2011.

and unloading. One of those pictures would become the yearbook cover photo of the 2011 *National Fisherman* magazine.

One of my friends, a lawyer from Rockport, Massachusetts, had offered to sue the State of New Hampshire for negligence for my fall off the pier. He was not interested in money and would only take a cut if he recovered some damages. Numerous politicians from the state had toured the pier after word of my fall reached the press, and the legislature procured money to replace the wall, hoists, and floats. This was a big win for the fishing community but hurt my case as the offending hoist was removed before anyone photographed the pier. At trial in 2012, the judge excluded almost all our photographic evidence as well as our expert witness as prejudicial. Also, fishermen who showed up to testify about the situation were severely restricted in what they could stipulate. The state's case was a version of blame the victim which cast sufficient doubt for the state to mostly prevail. I was found 80 percent at fault and the bar for recovery was 50 percent. This was my first taste of the difference between right and wrong versus legal and illegal.

Daniel returned to fish again in the summer of 2012. He would bring a computer and work on his thesis between tows. Observers always wondered about this strange boat that had a crewman working on his PhD. This year was hot, as it turned out the hottest year on record since the early 1950s. Why was this important, well, it soon became obvious to fishermen that the fish were distributed very differently than in years past. As part of the study fleet, I still had a bottom-temperature sensor, and as the bottom water temperature crossed fifty-two degrees, the fish left the nearshore area and the shallow water of Jeffreys Ledge and began to inhabit the very deepest parts of the western Gulf of Maine where water temperatures were still in the high 40s. Whiting was still present but had been joined by unprecedented numbers of butterfish, which are a mid-Atlantic species. Fishermen in the Gulf of Maine could only possess incidental quantities of butterfish, because they were not permitted for more, which led to enormous waste. My friend Tom with the *Midnight Sun* had a limited-access permit and landed large quantities of these fish, prized in Asian markets. Also, we noticed large numbers of squid. Again, we had the wrong permits and fed many to the birds. Fishermen did not then understand the damage to Gulf of Maine species that was occurring from this mid-Atlantic invasion but would learn in a few years' time. I consulted landings records as well as temperature records kept at a lab in Boothbay, Maine. I soon learned that in the 1950s cod, yellowtail flounder, and northern shrimp virtually disappeared. Haddock, redfish and pollack became the dominant species until the water temperature regime shifted again in the early 1960s. The difference was that in the 1950s there was no NMFS, no Magnuson, and no requirement that all species be at maximum sustainable yield simultaneously. People just targeted what was here to catch. That would not be the case this time.

Also, I began work on collecting fish samples for a new cod study. Initially a pilot project, the goal was to collect tiny fin-clip samples and analyze their DNA. The working hypothesis was that different cod stocks would be genetically different. Some previous work had been done on silver hake using samples from Nova Scotia to Maryland. That study concluded that adjacent samples had little genetic differences, but over the entire range, fish from Nova Scotia were genetically different from those off Maryland. This work, which was a logical outgrowth from our cod-tagging work would hopefully decisively prove that cod-stock boundaries were not correct, and, in turn, force assessment and management change.

David Goethel running the hoist controls lifting two thousand pounds of fish, 2010.

While regime change in the Gulf of Maine was occurring and fishermen were trying to understand the consequences, another massive shift was occurring in NMFS. The regional administrator abruptly resigned and retired and was replaced with a former one-term mayor of New Bedford. I was skeptical of the change for two reasons. First, New Bedford was a large-boat port. Second, I had been part of an advisory panel for a boat-and-permit buyback program he had administered in the late 1990s. There was no need for the advisory panel, as this man who had never fished knew everything there was to know about fisheries. This firsthand knowledge, combined with the fact he had lost reelection in his second term as mayor, set off alarm bells. However, my third and final term on the council would end in a year, and I would not have to deal with whatever issues cropped up. By the November council meeting, change was clearly in the air. Historically, the administrator had participated in council discussions, usually only pointing out obliquely when issues the council was considering might not be approvable by NMFS for legal reasons. On final votes, the administrator would usually abstain, citing the fact that NMFS had to approve the final measures or not. But not this man; not only did he vote, but he also actively berated council members who often had well-founded arguments against his point of view. Early on, we noticed he particularly browbeat women on the council, especially a woman from New Bedford. She was feisty and gave as good as she got but did not reapply for her seat when her term was up. About this time, another council issue popped up. Technology had overtaken the process and many council members now had both laptop computers and cellphones. Council binders had always been printed on paper and people not at the table communicated with council members at the microphone. All comments were recorded and available for later inspection. The head of NOAA had ordered the councils to go electronic to save trees, but this had unintended negative consequences. Audience members began emailing council members motions and snarky comments, and NMFS staff communicated with each other on erasable nonpublic sites in violation of right-to-know laws. How did I learn all this? Council members are seated very close together around a horseshoe-shaped conference table. The staff placed name plates randomly, so people of like mind are not necessarily able to sit together. Over the years, with the advent of computers, I had watched council members in session read newspapers online, including one in French, shop on the home-shopping channel, and, in the case of one NMFS employee, read *Golf Digest*. This usually went on during public comment sections of the meeting. My feeling was and still is that our job is to listen to public comment. Many people traveled great distances to make comments and would be appalled to know *Golf Digest* is more important than what they had come to communicate. However, council members receiving motion wording from lobbyists in the audience or snarky comments about another council member with whom they might disagree was beyond the pale. I approached the executive director to request that only staff members giving presentations be allowed to have computers and or phones at the table. This was duly debated by the council's executive committee, and I lost. Computers were here to stay, another wound to struggling democracy.

On a dark night in February, New Hampshire convened its town-hall-style meeting to select three candidates for the governor to submit for the NH council seat. New Hampshire is unique in that interested citizens, commercial and recreational fishermen, as well as other interested persons submit names and then ask questions of the candidates before ranking them for submission to the governor. So far, the governor has never added names of his or her own preference to the list. On this night, the assembled selected my wife Ellen as their first choice. Ellen is an invertebrate biologist who has taught introduction to marine science in school systems throughout the Northeast and who was about to open a small, summer museum on Hampton Beach. She was supported by both the governor and the state's congressional delegation and would be appointed to take over my vacated seat in August. The year's fireworks were just getting started though. At the end of February, NOAA head Jane Lubchenco would resign her position. Apparently, she had rankled enough feathers in Washington to be deposed, but the damage done by the sector management she championed would last the rest of my fishing career. In July, Senator Kelly Ayotte held a hearing of the Senate Commerce Subcommittee. Among those summoned was New England Regional Administrator John Bullard. The topic, the placement and cost of observers on fishing boats. NMFS was warning ominously, that absent increased funding, fishermen in New England would have to assume the cost of paying for observers. Senator Ayotte's previous position had been attorney general for the state of New Hampshire where she was known for a take-no-prisoners cross-examination style which had won her many cases. What Mr. Bullard did not know was that his personality, combined with my thorough briefing of the senator's staff, along with Senator Ayotte's thorough understanding of Magnuson, would soon lead to an epic clash. Among Mr. Bullard's statements was that the law requires fishermen to pay for observers, where in fact, the law says boats must carry observers if requested, must have a valid Coast Guard Safety decal, and must brief observers on the location of safety equipment. The law is silent on payment. There are cost-recovery provisions for limited access privilege programs, but the government had spent the last four years arguing sectors were not a LAPP. Senator Ayotte gave him a chance to amend his statement, but when he did not, she eviscerated his lack of knowledge of the law he was supposed to be upholding. The fishermen had won a battle, but, as always, we would soon lose the war.

In late August, Dan and I would have a long talk about both our futures. He was nearly through writing his thesis, and, barring some totally unforeseen event, would obtain his degree in stock-assessment science. The problem was that Dan still had the fishy gene and continued finding ways to return home to fish for three months or more each year. I wanted to know if Daniel wanted the

Left to right, David Goethel, Senator Ayotte, and Capt. Erik Anderson discussing fisheries policies at the dock of the Yankee Coop. Seabrook, NH, 2014.

boat and fishing permits. I was nearly sixty, and this was a young man's game. What followed was well-reasoned, as always. If I gave Dan the boat and permits, he would have to support Ellen and me for the rest of our lives, as the value of the boat and permits was our retirement fund. If he purchased the items at fair market value, he would have an enormous loan payment for the foreseeable future. So, either way, Dan would have little money. Dan had decided to take a job performing assessments and try to fix the process from within, to, as he put it, make sure no fishermen suffered from inadequate science in the future. Eric's famous statement, "Why would anyone do the job you do and put up with the crap you take for the money you make?" made clear Eric wanted no part of a fishing career. The combination meant I would fish until I no longer could, and the business would die with me.

Ellen had no sooner settled into her council seat than a surprise announcement was made at the November council meeting. The government had done an unannounced stock assessment of Gulf of Maine cod and determined that the population had dropped by 90 percent since their last assessment. How could

90 percent of a fish stock that was thought nearly rebuilt two years ago simply disappear? There were several possible answers: the fish had moved to deep water where few survey tows were made or had died trying to seek cooler water, or the government had grossly overestimated the stock size and by letting large boats catch large amounts of fish, albeit legally, under sectors management had decimated the stock. Or maybe the new assessment was wrong. Or maybe all the above. Whichever was the case, the cod-assessment scientist who was not trained in biology but had a math background dug in his heals on being correct, and the regional administrator who was dying to use his emergency action powers dug in his heals alongside him. No one was given time to investigate the issue.

The meeting immediately devolved into heated accusations from many people. The administrator claimed this situation was just like Newfoundland, where scientists in papers he knew of warned authorities to radically scale back fishing. Except it wasn't the same at all. The inshore, small-boat fishermen had been the ones to warn Canadian authorities that the newer, larger, ice-breaker-class boats that fished through the ice on spawning aggregations were the issue, but the scientists stubbornly clung to their projections, citing the high-catch-per-unit effort of the new boats. Further statements on his part about government scientists never being wrong and fishermen just catching the last few huddled masses of cod set the tone for the rest of the day. As soon as the news had been given out, Ellen called me to let me know what was happening. I grabbed a list of cod papers that I kept in a computer file on various cod subjects. These included stock-boundary issues, causes of the Newfoundland cod collapse, and fish aggregating tendencies, as well as data on cod's thermal tolerance. When I arrived, the council had already voted down two options and was in the process of eliminating their last possibility. The coalition's director briefed me on all the various charges, counter charges, and foolish statements that had been made. It would soon be my turn not to have my finest hour. Because I had not spoken on the first two motions, I was recognized to be one of the limited commenters for the last motion. Rather than call the regional administrator out by name, I started with the fact that numerous erroneous statements had been made about scientific papers and outcomes that morning. I implored people to talk less about science and instead to read the papers. Then I offered a list of seven papers which might help them understand what they were dealing with.

I thought I had couched my criticism in neutral terms, but the regional administrator turned red and demanded the chair recognize him. The chair tried to move on, but Mr. Bullard activated his microphone and fairly screamed that he didn't have to read scientific papers because his scientists read them for him and they had told him there were no cod. Suffering my own bout of anger management, I replied that those must be the same scientists who were off by 90 percent the last time, as I walked back to my seat. The final motion would be voted down. The regional administrator would announce an emergency action effective in three days. In that period, the fleet would land two million pounds of cod as people tried to use up their legal allocations. Most of those fish were caught by large boats fishing on Middle Bank because the weather had turned too dangerous for the small boats. In all my years of being involved in this process, I never witnessed such abject stupidity as just then. The outcome was entirely predictable and entirely preventable. An entire school of cod in Massachusetts Bay was slaughtered for absolutely nothing because one egotistical fool had to prove how powerful he was. I am not sure we had a cod problem of epic proportions before November 11, 2013, but we damned sure had one afterwards. Also, I had been wrong about sectors doing lasting damage to fish stocks in five years—it had only taken three, and the last three days of those three years were a classic example of the destructive fishing power of large vessels.

CHAPTER 32

You Lie for a Living!

My crewman Mike (who was former crewman Jeff's younger brother) and I needed a new game plan. Not only were we limited to two hundred pounds of cod per day, but should we catch more, they would be deducted from our meager quota. We settled on a two-prong plan: fish for bait skates and lobsters in the fall, and when the skates dispersed in early winter, head out to the deep water of western Jeffreys Ledge to fish for lobster, which rose substantially in price in the late winter. The ASMFC announced the shrimp season would only be six weeks, from mid-January to the end of February, because shrimp populations were declining rapidly. That was not a big surprise, as the winters had been warm for longer than the five-year life of a shrimp, and recruitment was poor because it is dependent on cold, bottom-water temperatures in January and February. I knew bottom-water temperatures had been climbing because I had been given the study-fleet, bottom-water temperature sensor. Still, the sight that greeted Mike and me in pre-Christmas December while towing in a heavy snowstorm off western Jeffreys gave both of us pause: A brown pelican landed on the rail to beg for food. The bird must have strayed north in the summer and was now trying to fly south for warmer weather. It was NOT impressed with the current weather. Mike's father was an avid birder, so Mike snapped a bunch of pictures with his cellphone and sent

"What do you mean it's going to snow? Where am I anyway?" A brown pelican makes a wrong turn at Cape Hatteras and has an emergency landing on the *Ellen Diane* during a snowstorm, to fuel up on the boat's catch. December 2, 2014.

them back to his dad when we had service. He received an excited email telling him a brown pelican had not been seen in New Hampshire in over one hundred years! The bird was a picky eater, scarfing down a small cod and a couple monkfish, but it turned up its beak at the rest of the fish we offered. It sat on the rail for almost an hour watching Mike measure and band lobsters and clean fish before flying off into the snow never to be seen again.

In early 2014, I attended a cod-stock symposium where work I was involved in, as well as that of other researchers, was presented and reviewed. I was surprised to find nearly one hundred researchers in attendance. Here is where the concept of two spawning groups of cod, those that spawn in the spring south of the Isles of Shoals in Ipswich Bay and those that spawn in the winter off Salem, Massachusetts, in Massachusetts Bay was officially unveiled. DNA showed these to be two genetically distinct groups of fish that were also different from those on Georges Bank. Scientific meetings adopt change by consensus, and while nearly everyone agreed, three people did not, saying a bigger sample size and wider geographic representation was needed. I was discouraged, but the researchers I worked with decided to put forward a comprehensive program to sample fish from the Canadian Scotian Shelf to the Jersey Shore. They also archived historical samples if they contained verifiable position information. When this was completed, no one could downplay the results.

My son Daniel received his PhD in May and already had lined up a job working for NMFS at the southeast region in Miami. Numerous scientists in New England asked him to work here, but he refused. He had watched how the sons of fishermen were treated and wanted to go to an area where no one knew my name. Tom's brother, Salvatore, whose lab I had used at the aquarium, would ultimately end up in the regional office. He was ostracized by the scientists as a plant for the fishermen and shunned by his family who considered him allied with the government. I had witnessed how the government distrusted him firsthand. He had an office in the post office with other NMFS employees in Gloucester at the time. I stopped by to invite Sal to lunch. As we left the building, he yanked me into an alley and behind a dumpster. I asked him if he was crazy, but, a minute later, another NMFS employee came out scanning the street for the two of us. Sal said this happened every time a fishermen stopped by, until he realized the significance of the shadow, who would take an adjoining table to eavesdrop on the conversation. Daniel had heard these and other stories and decided never to work in the Northeast. So, Dan was off to Miami. I was incredibly sad, as having him on the boat had been some of the best fishing memories of my life. We had experienced every high and low moment small-boat fishermen can encounter. Record hauls, small hauls, unforecast storms, entanglement in uncharted wrecks and hangs; you name it, we had done it all and come home victorious.

The council was preparing a framework to institutionalize the radically lower cod quotas recommended by the latest cod assessment. Ellen queried me after returning from a council meeting as to whether I thought there was a good-old-boy network at the council. I had never really thought about it, maybe because I was an old boy, but I answered with a question, "Why, what is happening?" She replied that the staff, both NMFS and council, treated her like an equal, but many of the men on the council treated her like she did not exist or, worse, like she did not belong there. She was having problems with the regional administrator, who came over to her chair before a meeting while she was seated, leaned down on the table, pointed his finger in her face, and lectured her about being my mouthpiece, not upholding her oath of office, and other such similar statements. I tried to joke, saying look how many of your motions I have spoken against, but this was no laughing matter. My wife is an outstanding biologist, and there is no one right answer in fishery management. To be lectured about biology by a guy who had a degree in architecture and had three staff constantly seated around him to explain biology because she brought an independent point of view was a clear example of just how far the process had jumped the tracks. My wife asked both the executive director of the council and the council chair to come stand beside her if they witnessed this behavior. After the executive director stopped by a couple times, the problem stopped.

The final vote on the revised quotas would be taken in November, with the number being two hundred tons for each of the next three years. To put this in context, I brought a picture of a research half-hour tow that had captured over a ton of cod the previous week. Doing the math, I pointed out I could catch the entire cod quota in the Gulf of Maine in a week and not break a sweat. There was one bright spot in this mess, however. The framework proposed to open an area north of Rockport that was currently closed to give us a place to fish for flounder in June. Also, fishermen were beginning to see rapid increases in the amount of haddock present. Just like in the 1950s, haddock, redfish, and pollack were doing very well. The 2011–13-year classes of haddock were among the largest ever recorded. History really does repeat itself. Shortly after this framework passed the council, the ASMFC held its annual shrimp meeting and announced there would be no season due to record-low biomass. This was going to be a long winter. Mike had left to take a job on a loading dock at his brother's employer. I was back advertising for crew again. With no shrimp season, Ellen and I decided to take a vacation and visit Dan in his new home in Miami.

Dan had settled in an older part of Miami, which by South Florida standards was relatively safe as well as close to work. He had rented a house with his girlfriend, and they seemed to be enjoying life in the neighborhood. Daniel had learned there was a drunk in the neighborhood who stole packages out of mailboxes and off porches. His solution was to buy the man a bottle of rum every week as protection and ask him to guard his mailbox. This worked for several years. The guy would chase off other miscreants who considered raiding his neighborhood. Unfortunately, we had to return to New England and the unseasonable cold and extreme snow that had been going on for several weeks. I found the boat entombed in ice in the marina slip. Worse, the thru-hull fitting for the washdown system was frozen, something that had never happened in all the time I had owned the boat. If the fitting cracked from the ice expansion, the boat would sink. Temperatures remained below zero, and the ice in the marina continued to thicken. Oak pilings, fifteen inches in diameter, snapped like toothpicks as the tide rose and fell. One crashed onto the bow of my boat. The only time the temperature rose was when it snowed, and it snowed a lot. Ellen snowshoed into our backyard and onto the roof of the garage to remove the weight from the buckling roof. The snow was deeper than she was tall. Meanwhile, the cod kept swimming south in search of warmer water out of the closed areas in Massachusetts Bay. Most boats were iced in like me, but a few large boats followed the fish into the Great South Channel off Cape Cod where they were now magically Georges Bank cod and subject to a much higher quota. The fish kept going, ultimately ending up south of Block Island, Rhode Island. My fellow researchers asked the State of Rhode Island to procure some samples for DNA analysis. Those fish ultimately stayed in that area, recolonizing southern New England waters.

I finally broke out of the ice on the last day of February. The marina was a shambles and would take all spring to rebuild. As always, when water temperatures get very low, the fish depart for warmer waters, but the lobsters stay on the western side of Jeffreys. Our first day out, we caught our legal limit, which sold for a stratospheric price because no one else could haul gear. We also caught some nice haddock. Maybe we could do something in March before the area shut down. Also, Ellen had been renominated for a second term on the council by the state as the preferred candidate for New Hampshire from the three names submitted to the secretary of commerce.

However, we both felt the chances of reappointment were slim. Indeed, in late June, Ellen would learn she had not been reappointed to the council. This sparked an angry exchange between the head of the marine division of New Hampshire Fish and Game and Regional Administrator John Bullard, who was appointed by President Obama and who claimed Ellen violated her oath of office. Apparently disagreeing with the regional administrator was now a violation. The man lied as effortlessly as ducks swim through water. Our head of the marine division is one of the quietest, most thoughtful people I know, and to have a public confrontation of this magnitude indicated the depth of anger he was expressing.

By late summer, it appeared increasingly likely NMFS would try to shift the cost of observers to the boats. Small boats simply could not absorb this cost of seven hundred dollars per day. Now, I have always been hard on this program, because it had strayed so far from its roots in cooperative research in the mid-1990s. As originally conceived, it was supposed to be a low-cost alternative to sending highly paid, PhD, science-center scientists to sea to gather biological information. Biology graduates would weigh discards and take various biological measurements on 4–7 percent of the trips. They got on the boat, did their job, and went home. Then the government got involved. At that time, Regional Administrator Andy Rosenberg created a separate branch within the science center to collect the data and check for accuracy. The observers could not be government employees for several reasons, so requests for third-party, for-profit contractors were put out. Coincidentally, Andy Rosenberg would become the president of the largest of those companies. At every step of the way, more information was required, more bureaucracy both in the companies and NMFS was required, and no one cared because the taxpayer was footing the bill. So, a program that had an original cost of about two hundred dollars per day had morphed into a seven-hundred-dollar-per-day beast, and that was only the "direct" costs. Factor in "indirect" costs and you were up to eleven to fourteen hundred dollars per day. The government started out with a budget Yugo and ended up with a stretch limousine with a hot tub. But I think what galled fishermen the most is that observer training concentrated far too much on teaching these people that fishermen were the enemy: that fishermen cheated, they took fish illegally, they would impede you from doing your job, and a host of other patently ridiculous statements. What no one checked was if the observers got seasick or whether if they had basic boat sense and training to obey the captain's orders in matters of moving about the vessel safely. Observers were told that they could not help the crew, even though the observers' work made the crews' jobs far more difficult and lengthier.

How do I know all this? Once most observers learned I had a biology degree, years of experience on the water, and would not eat them for lunch, they told me these things and a lot more. Many bits of information that were collected, like the daily safety inspections—held even though our equipment is certified by the Coast Guard—were not instituted for safety but so the observer could establish dominance over the captain and crew. But the observers took it from both sides. They were constantly debriefed by NMFS federal enforcement agents looking for evidence of collusion or fraternization. Observers could only go on a boat twice a month, a limit meant to keep them from getting friendly with the fishermen. Despite the government-inspired paranoia and harassment, some people liked the job and became quite proficient at it. These were the men and women boats hoped to have assigned to them. Most kept lists of the expiration dates of safety equipment so they could fill out their forms without holding up the boat in the morning. They could efficiently sort and weigh fish without mistakes. If they finished before the crew, they would offer to rinse or box fish, little things that made everybody's day easier. I never saw experienced observers falsify data or offer to let you discard fish illegally. I did see newbies misidentify fish and fill out paperwork that was made up because they were rolling around deathly seasick, and once, I caught a man misreporting red hake as white hake because he had been told to find white hake by a supervisor. I made sure these people were fired.

Against this backdrop, all fishermen and fishing organizations began a full-court press. Letters were written to all congressional delegations. Associations and organizations wrote letters to the editor, op-eds, and offered to take out television crews. We explained the problem in a way non-seafarers could relate to. Imagine having someone show up at your house every morning and jack up your car to inspect the brakes, even though you have a safety sticker, and then ride along in your car to make sure you do not exceed the speed limit on the interstate, and then you must pay for it. The same ENGOs that are calling for this would have the American Civil Liberties Union on speed dial if it happened to them. I was writing seven or more letters per week. Our organizations were fielding calls for interviews, and camera crews wanted to film the situation. I received requests for interviews from all over the country, places like Utah, Colorado, and California. I didn't know if any of this would make a difference, but we would not go quietly.

Against this fevered background, Senator Kelly Ayotte held a field hearing on the state of New Hampshire commercial and recreational fishing in Portsmouth, New Hampshire, in September of 2015. Just before the date, I was interviewed by the *Portsmouth Herald* about fishing problems for both commercial and recreational fishermen off New Hampshire. It was pretty much the usual stuff about how you cannot run a recreational or commercial fishing business if you cannot catch and keep any fish. On top of that, commercial boats would soon need to pay for observers at seven hundred dollars a day. If the interview had ended there, all would have been routine. The reporter pressed as to why I thought this was happening, and because I was tired, I answered honestly, not diplomatically, "NMFS lies for a living!" I almost asked him to keep that deep background, but I was tired and fed up with the foolishness of the last two years. Well, of course, that ended up being the headline.

Senator Ayotte had asked me to represent both groups of fishermen and to ask questions on their behalf. Depending on the response, she would follow up with further questions. Naturally, the room was packed with both fishermen and reporters, as well as with television cameras. The table was a horseshoe with John Bullard and the head of the science center, Bill Karp, on one side, Senator

Ayotte, her staff, and other congressional staff at the head, and me and an owner of a fleet of party boats on the other side. Bill Karp is a nice man who had tried to correct some of the endemic structural problems within the science center during his time as director. Since much of what everyone wanted to know was about the status of cod, the first couple questions and answers were about science and were a discussion between Bill and myself. The discussion was cordial, although we clearly disagreed. Then came a question on why recreational vessels could not keep even one cod. Rather than answer the question, John Bullard reached in his pocket and pulled out the newspaper article. The room fell silent as he began by asking, "I want to know if you were quoted accurately in this article." "Yes, I was." Bill Karp tried to put his hand on Bullard's arm like "don't do it," but it was brushed off. Bullard continued, "You have not got the guts to come over here and say that to my face!" I shot Ellen a glance in the audience and was getting the wife stare equivalent of "don't do it." I decided on a compromise statement to save Senator Ayotte from embarrassment. I responded, "I can walk over there if you like, but that would probably be disruptive, so I will say it loud enough that everyone can hear: You lie for a living!" Cameras clicked, television footage was recorded, people buzzed, and for a minute, I figured all hell would break loose. Instead, Senator Ayotte chastised Bullard for not answering the question and breaking decorum as a government official. After the meeting was over, both officials rose and exited the building before the press could reach them. Senator Ayotte and I fielded a few questions until finally everyone except Ellen and the senator's staff had departed. At that time, the senator quietly thanked me for speaking for the fishermen and keeping my cool under fire. My response was, "I think this illustrates how deep the rift is between the fishing community and NMFS, and maybe some good will come of it."

CHAPTER 33

The Crazy Premium

As the November 1 deadline to start paying for observers loomed large on the horizon; fishermen, sectors, and fishing organizations had to decide how to proceed. Filing a lawsuit with a request for summary judgement was an option, but how would we collectively raise the money? Our usual fund-raising consisted of bake sales, raffles, and fish fries, and would not even scratch the surface of the money needed. We had the public on our side, and NMFS had a black eye, but in a couple days, the story would fall out of the news cycle, and the bureaucracy would lumber on. Our sector manager came up with a novel payment plan. Since the sectors would have to pay the bill (NMFS had decided chasing fishermen for the money and suspending permits would be both costly and very bad public relations) on behalf of the fishermen in the sector on a quarterly basis, the sector would bill members on a pooled coverage basis. For example, if the sector had one hundred observed trips at $700 per trip, the total bill would be $70,000. If ten boats fished, each boat would pay $7,000. This was still a huge amount of money but would not be as debilitating as one boat having fifty observed trips and paying $35,000. This had happened to me, where I had been "randomly" selected forty-seven days in a row.

Late in November, I received a phone call from a lawyer representing the group Cause of Action Institute (COA) in Washington, DC. The lawyer wished to fly to New Hampshire to interview me and determine if I might be a suitable plaintiff in a lawsuit against NMFS (actually against the cabinet-level agency head). Cause of Action in law is defined as facts that enable a person to bring action against another. Basically, the interview would determine if I could meet that burden so that a suit could be filed on my behalf. The institute would pay all the bills; I would need to be present for the trial, do interviews, and read the briefs for accuracy. Cause of Action would pursue a judgement in the fishermen's favor all the way to the Supreme Court if necessary. My response was, "How soon can you be here, and do you need a ride from the airport?" After an all-day interview in my living room, which included the legal aspects such as was I actively fishing throughout the period, a lengthy, detailed description of the nuances of the myriad actions that had been enacted since Amendment 16, and a complete dissection of my personal life to make sure there would be no character surprises, the attorney returned to Washington to consult with superiors. Three days later, I was again contacted with a request to fly to Washington for two days of consultation. On arrival, I was transferred to COA headquarters where I met with four attorneys who would be my legal team. I also met the president of the organization, who asked extremely pointed questions. This was a donor-funded group that handled numerous cases against the government simultaneously. Millions of dollars were in play, and the donors did not like mistakes or surprises. After that, the five of us got down to work. I am not a lawyer and did not understand much of legal back and forth around the table. The attorneys spouted case law the way I recite LORAN numbers for wrecks. While there would be numerous charges in the brief, the core issue was the fact that NMFS was not a taxing authority but was imposing an illegal tax on the fishermen. To support their case, COA needed detailed histories of discussions. To that end, they filed Freedom of Information Act (FOIA) requests against every government official remotely connected to Amendment 16 and subsequent actions, as well as against all council members and staff. From this time on, a COA attorney would attend all council meetings to take notes and observe the proceedings. Government officials are required to preserve emails and phone logs. The council had been briefed about needing to have separate email accounts for council business, but many people did not do this. By the time I got home, I had numerous emails directing me to preserve records as well as a few asking me what the hell I was up to. Apparently, many council members had been lazy about keeping their emails separate from their personal stuff, some state officials used state servers, and at least one person had used an academic server. Further, government officials had communicated using personal emails, either knowingly or unknowingly, to avoid federal record-keeping requirements. COA was already in court with Department of Justice lawyers over the FOIA's scope. There was a new sheriff in town, and this one took no prisoners.

The money ran out for monitors in March of 2016, and the government instituted its payment plan. Shortly after this, a COA attorney witnessed a

NMFS employee emailing the regional administrator at the April council meeting. Another FOIA was filed, but no records were located because the website apparently erased after a few minutes. Meanwhile, I had been tasked with writing op-eds explaining the problem. COA had a public-relations team with access to the largest newspapers in the country. The rules were simple: write nine hundred words or less and make sure it was understandable to readers. I would submit these articles to COA, they would edit them slightly, mostly to simplify the grammar and content, and then submit the op-eds on my behalf. One of the most popular stories appeared in the *Wall Street Journal*. I received calls from all over the country with offers of support, including one call from my uncle in California urging me to stay the course. If our case was heard in the court of public opinion, we would win hands down, but, instead, it would be heard by Judge Joseph LaPlante in Concord, NH, district court. The case began early in the morning, and I was on the stand for most of the morning session. I had assumed NMFS counsel would try the case, but it was presented by Justice Department attorneys from the environmental law division. There were some preliminary theatrics as government attorneys tried to introduce documents not turned over in discovery, but finally we got started. There was a through-the-looking-glass quality to my time on the stand because the judge and the government lawyers did not understand much of the terminology. Ordinarily, that would be a problem for the attorneys to sort out, but in the interest of moving on, I would ask the judge for permission to explain the terminology in lay terms. The afternoon session was spent on the legal aspects of the case, and then it was my turn to have little understanding of the back and forth. Around 3:00 p.m., the judge said he had heard enough and would render a decision in the coming weeks.

The verdict was announced in a written ruling on July 29, 2016; we lost on all counts entirely because of one overarching legal issue: we were time barred from suing. According to the law, we should have sued within thirty days of publication of the final rule for Amendment 16 in 2010, not within thirty days of the announcement that fishermen would have to pay in 2015. This still makes no sense to me; how can you sue a hypothetical and what lawyer would have pursued the case? Now, fishermen had to add clairvoyant to their list of talents. This was the second time in my life legal and illegal trumped right and wrong.

The COA attorneys were disappointed, but immediately filed an appeal with the circuit court of appeals in Boston.

The Fifth Circuit Court of Appeals ruled against us again on the time bar issue on April 14, 2017. However, much of the judicial panel wrote in its conclusion: "However given NOAA's own study which indicated that the groundfish sector could face serious difficulties as a result of the industry funding requirement, we note this may be a situation where further clarification from Congress would be helpful . . . " Basically, had we not been time barred there could have been a ruling that NMFS consult Congress. Based on that tantalizing quote, COA appealed to the US Supreme Court. The court refused to hear the case in 2018. It was over, or not, as FOIA requests are still being litigated in the DC district court. Time and taxpayer money are always on the government's side. There is absolutely no reason for the government to compromise. Absent groups like COA, no citizen can possibly hope to achieve legal redress against their government. At least my two sons and uncle were proud of me and rightfully pointed out I was an asterisk in history with my name on a Supreme Court case.

Always ready to help out, the backup crew clearing the deck with a smile. Ellen Goethel, December 2017.

My wife was trying to ease me into retirement, but I still enjoyed fishing, despite the impediments. My friends were all retiring, and the ocean was becoming lonely. Tom was in the early stages of turning the *Midnight Sun* over to his son, Tom Jr. Red was now manager of the coop but told the board to advertise for a manager. He would do the maintenance, but the government reporting and constant carping about prices from some fishermen were causing his health to head south. He was about to have yet another stent put in. Erik was still lobstering, but only during the summer season. In short, there was a generational shift occurring in the fisheries. Our oldest son, Eric, owned a condo in Stuart, Florida, that he commuted to seasonally from his tugboat job in Boston. In classic Eric form, he wanted us to buy a place close to him, so if he had to put us in the home (rest home), we would be nearby. He meant this as a joke, but still . . . At any rate, Ellen went down, toured a few places, and found one she really liked. The location was west of Route 1, in a quiet area and fully built out. The price was about a third of a similar unit in Hampton. I told Ellen to offer 10 percent below the asking price and was surprised when the seller accepted the offer. I guessed I would be wintering in Florida the next year. There had been no shrimp season for the last three years, and, given water temperatures, I assumed there would not be any soon.

I had one last trip through the court system to take though. I had been sued by a crewman for an injury. The man could not prove it happened on the boat, but I could not disprove it either. The case languished for several years, as every time it got close to trial, the man fired his attorney. My insurance company had learned that the address and social security number he had furnished were both false, the IRS was looking for the plaintiff, and he would not show up for a deposition because he lived in another state. I wanted this case closed, because I wasn't sure I could sell the boat while it was outstanding. Fed up, I left for my new winter home to try recreational fishing on the shores of sunny Florida.

Surf casting with Eric in Florida was quite enjoyable, and I soon got the hang of it. There are many species of fish in Florida, and part of the excitement was figuring out what was on the end of the line. The target species was pompano, which were very good eating and fought well, but you could easily catch a dozen different species in every outing. Two things to avoid were lionfish, an invasive species, and saltwater catfish, which had a spine tipped with venom which would not kill you, only make you wish you were dead. Other than that, everything was good eating. I received a call in January announcing that John Bullard had retired. Hallelujah, maybe his replacement would let the council do its job. In mid-March, I heard from my insurance company that a trial was scheduled for early May with Judge Joseph LaPlante. I told the insurance company I had appeared before him with COA and asked that they check to make sure he would not have to recuse himself. They thanked me and phoned back several days later that it would not be an issue. The hearing would be an attempt at a negotiated settlement; a trial, if necessary, would occur later. I showed up on the appointed day at the given hour and had to sit through a most bizarre hearing. The ex-crewman began by demanding the bailiffs be arrested because they had searched his car. Next, he was having a difficult time following the proceedings because he had not taken his medications. When the judge asked what medication that might be, he replied, "One hundred mgs of oxycodone every four hours." The judge looked down his glasses and asked if he would like a glass of water. What followed was a bizarre and convoluted tale full of accusations and falsehoods. I was told by the insurance company attorneys to just grin and bear it. When they were finally allowed to speak, the ex-crewmen interrupted them constantly. Finally, the crewman's attorney was told to control his client, or he would be removed from the room. This guy was odd when he worked for me, but now he was one side or the other of crazy. Around lunchtime, the judge put each group in a separate room and planned to do shuttle diplomacy. I can only assume what he told the other room, presumably to get serious about accepting a small settlement. I do know what we heard. First, the judge said he remembered me and was truly sorry if his ruling had made it necessary for me to hire people like the plaintiff. Next, he told the attorneys that they would need to make a higher offer than usual to pay the "crazy" premium because there was no way the plaintiff would make it through a trial without disrupting the court and causing a mistrial. I reminded the judge there was an IRS garnishment outstanding. I was asked to leave the room while the judge and attorneys talked figures. My assumption was that the number to be offered was the garnishment plus travel compensation. Whatever it was, judging by the screaming you could hear through the walls, it was not well received. We were asked to provide a bailiff with funds and a lunch order for subs and tonic. While eating our lunch, the judge reappeared to say the settlement had been agreed to and the plaintiff escorted to his car by bailiffs. We were strongly advised to take our time finishing lunch so the plaintiff would hopefully be long gone. The judge was clearly worried about a confrontation.

Crew, as always, was still an issue. I was consistently carrying two people now in the hopes one would show up on any given day. I fired one in 2017 for fentanyl use. He called my wife saying he was sorry for not showing up, but he could not work until he got his check so he could buy some more fentanyl. My wife called me stunned. I called the crewman to tell him anytime he had a ride, he could pick up his raingear and boots. He showed up on payday and literally ran back to the car, which sped off to buy drugs. The next one arrived as a born again with a Bible and left as a felon with a jail sentence. The poor college man who spent the summer working with him tried to keep him straight, but the bigger the checks, the more trouble he got in. His replacement was his sister's boyfriend, who had just been released from state prison. This man was on parole and had apparently spent sufficient time in jail that he had no desire to return. He soon ditched the sister and took up with a schoolteacher. Nine months later, he quit to be a baby daddy.

Despite all the drama on the water, 2018 and 2019 both produced strong catches of haddock and whiting for good prices. During the Cause of Action case, Congress would not get involved in the observer debate and rightfully so. However, with our final defeat in 2018, dialogue resumed. Senator Ayotte had been defeated for re-election. Senator Shaheen was our senior senator and was assigned to be the ranking democrat on the Commerce, Justice, Science Appropriations Subcommittee of the Appropriations Committee, which oversees fishery legislation and NOAA's budget. Ellen and Senator Shaheen had a friendship going back to Ellen's time on the conservation commission when Senator Shaheen was then governor. We were both contacted to provide an assessment of the status of ground fishing in New Hampshire. The report was bleak: there were only six active boats. Three of those men were at retirement age. We were then asked how much money we thought would be necessary to cover observer costs for the region. My guess was two million, but I would have the coalition's representative in Washington provide an accurate number after canvassing the sectors. Meanwhile, the senator's staff contacted NMFS; their response between two and ten million! The coalition came back with a figure of $2.5 million. Senator Shaheen was on the appropriations committee and had been a governor of a state with no sales or income tax. She knew when she was being played by the bureaucracy. She secured ten million but put a footnote in the budget requiring a detailed spend plan from NMFS. The bureaucracy may have won the battle, but they secured a powerful watchdog in the senate for years to come.

CHAPTER 34

One Last Challenge

While most fishermen were preoccupied with the costs of observers and whether they could survive, two other sector-related issues were percolating in the background. The council, realizing it had not limited the ability of entities to accumulate excess shares of the groundfish fishery, and thus de facto control all fishermen's ability to fish, had begun work on Amendment 18. Most fishermen wanted a small number, such as 2–3 percent, but several big players owned considerably more than that amount and wanted to be grandfathered in or they would file suit. The council had spent several years debating various ideas, but ultimately was getting nowhere. They finally hired a consultant who produced a number 15.5 percent, which was slightly higher than what the largest permit holder then held. If the council adopted this number, seven entities could own all the groundfish rights in New England. While public hearings were being held on Amendment 18, the largest permit holder was arrested for fish laundering and tax evasion. That alone should have set off warning bells.

Carlos "The Codfather" Rafael was already well known to government officials before sectors. Outwardly charming, he seemed to delight in tweaking authority. He had been around the waterfront since the last failed experiment with hard quotas in the 1970s. If he did not spring corrupt from the womb, he had corruption down to an artform at an early age. As I tried to convince the council and NMFS during my term on the council, you need both a crooked fishermen and a crooked dealer to sell illegally caught fish. No one ever seemed to get the equation. Enforcement went after fishermen for the most minor offenses but turned a blind eye on dealer-laundered fish. In the 1970's, with quotas on only three species, fish would be unloaded as what they were, say, cod, and leave the dealer's shop as something else, say, pollack. Carlos dusted off this scheme with sectors, only now he was vertically integrated and controlled the paper trail from start to finish. But he still had a problem: his newfound pollack were worth fifty cents while cod were going for two dollars. What to do? Have a crooked New York City dealer pay him the difference in cash, which Carlos referred to as bags of "jingles," and have a crooked sheriff working at Logan Airport spirit the cash through customs and on to the Azores. Carlos's boats carried observers, filled out logbooks, and were observed by dockside monitors; everything ENGOs said was necessary to monitor catch. So why was he getting away with this? Well, all this paper went to different people in government, and as is so often the case, the left hand had no idea what the right hand was doing. Carlos's grip on the waterfront tightened, but he had one problem: his own big mouth. Carlos loved to talk and tweak the establishment. He seldom attended council meetings because rules did not matter, but in one rare appearance, he went to the microphone and stated, "I am a pirate, it's your job to catch me!" In another interview, while famously puffing on his perennial cigarette, he told a reporter that small boats were all through and would have to sell out to him. He famously referred to hundreds of fishermen as mosquitoes on the balls of an elephant. He was not popular outside New Bedford. While he may have been fooling NMFS, the IRS, who had already convicted him once, became very interested in how a man who owned 15 percent of the fishery was only posting a five-million-dollar gross. Posing as Russian mobsters looking for a way to launder money, they asked Carlos to sell his business, but only if he told them how to move the fish. Carlos laid out his scam in great and thorough detail, even saying once, if you guys are undercover cops, this will be some bad shit! Well, it was in February 2016 the IRS and NMFS seized his boats, permits, records, and fish house. Carlos was arrested along with his bookkeeper, some captains, and his fish dealer. Several days later, two corrupt law enforcement officials were rolled up. Carlos was grounded for the moment.

The government wanted to make an example out of Carlos. Enforcement had been stung by an inspector general report in 2010 which found "systemic, nationwide issues adversely affecting NOAA's ability to effectively carry out its mission of regulating the fishing industry . . . particularly in the Northeast Region." The inspector general went on, "We find it difficult to argue with those who view the process as arbitrary and in need of reform." The system used civil penalties administered by Coast Guard administrative law judges. The Coast Guard, of course, was the agency most often bringing cases to NOAA enforcement. In essence, the judge and the prosecutor were on the same team.

NOAA enforcement leveled increasingly large civil fines, which were paid to an asset-forfeiture fund run by NOAA enforcement. This money was used to buy things like an undercover yacht. The scandal forced a few resignations and a massive overhaul of a badly abused system. The agency licked its wounds and waited for a big break to get back in the game. Carlos was their big ticket. Instead of using the civil law system, the IRS and NOAA went for criminal charges. They wanted Carlos in jail and all his assets seized and sold. For once, fishermen and the government were on the same team. I do not know anyone on the water that did not want Carlos removed permanently from the fisheries. The tax-evasion case was pretty straight-forward and would net Carlos jail time, but seizing and selling his businesses ran afoul of the Eighth Amendment of the Constitution. The fines could not exceed the cost of the crime. Basically, the justice department had to prove the value of the fish he laundered and that was the extent of the assets that could be sold at auction. Meanwhile, NOAA was paying to keep his fleet in storage and afloat, which proved difficult and expensive. Carlos would go to jail for forty-six months, but only two of his boats and permits would be sold at auction to cover his crime. He would pay a three-million-dollar fine and sell his boats and permits and keep the money. He could not own boats again, but he was retiring anyway. Crime may not have paid, but it did not leave Carlos hurting for money. Last I heard, he was buying a New Bedford golf course.

Carlos would be officially free in March of 2021, but the damage and destruction left in his wake still haunts fishermen today. His legacy was a lasting view held by NMFS that all fishermen are crooks and must be watched constantly. Faulty science got a free pass, because it now had the excuse assessments were off because of misreported catch. The council's Amendment 18 was finalized in April of 2017, while Carlos was in jail. His boats and permits would be sold to a private equity firm, Bragel Capital, who ran their business under the name Blue Harvest Fisheries. This company hired all the right people and paid lip service to all the right statements but would soon fall under the investigative microscope on its own.

Having vanquished Cause of Action at the Supreme Court and proved all fishermen were guilty by association with Carlos Rafael, the council and NMFS began work on Amendment 21, officially known as the Omnibus Industry

Funded Monitoring Amendment. This amendment would be put in place in February 2020 guaranteeing, now or in the future, NMFS could force all fishermen to pay for observers. This spawned a whole new crop of lawsuits which are still being litigated today. During this same period, work was begun on Amendment 23 with a goal of 100 percent monitoring on all groundfish trips. The amendment is still not finalized, as Covid reared its ugly head and government work slowed to a crawl. However, should this be finalized and the courts not strike it down, fishermen will become the first Americans to be watched by government minders every working minute. The road to hell is paved with good intentions.

The haddock fishing in 2019 was outstanding; the best I had ever seen. Whiting and herring were excellent as well, but hardly anyone was left fishing. Lobster fishing was also thriving, and gear conflicts became more and more of an issue. A few new entrants in the lobster fishery tried to push the dayboat draggers out. They learned a lesson fixed-gear fishermen have learned repeatedly in my lifetime: mobile and fixed gear cannot coexist on the same bottom. One boat from Rockport proved particularly problematic. The captain was reputed by numerous Rockport fishermen to have serious drug issues and a big mouth, not a good combination. One foggy summer day, he had nearly rammed me while passed out at the wheel. New Hampshire Fish and Game had seized a boatload of his illegal traps, but, somehow, he was still in the fishery. Late in 2019, he started screaming at me on the radio, after apparently losing some traps. I was a dinosaur and he wished I would hurry up and die. Rockport would have the biggest party ever on news of my death. He made me laugh as he struggled to put together coherent statements. He was not the first drug-addled fool I had dealt with, and, unfortunately, probably not the last. All the earlier ones had ended up the same way, overdosed and dead. The only question was when, not if. Since the theme seemed to be loosely based on evolutionary biology, I decided to go with it. I pointed out that as a young lobsterman he was an emerging mammal, and the dinosaurs ate an awful lot of emerging mammals before going extinct. I, not so politely, pointed out it would be in his self-interest to stay out of the dragger bottom. Apparently, the fleet thought my analogy humorous, because I was widely congratulated for putting the young fool in his place when I got back to port. Nevertheless, these encounters made the fall extremely tense, and I seriously considered putting a firearm on the boat for protection, as law enforcement seemed to have no appetite for dealing with the issue.

Before leaving for Florida in January 2020, I told the coop board of directors and manager that I was putting my boat up for sale for the purposes of retiring. Since I had been their largest producer the previous year, this caused great consternation. I was strongly asked to reconsider and fish one more year while they tried to reconfigure their business model. Senator Shaheen, working with the coalition and my wife and me, had secured another year of observer funding, and my health was good, so I decided to fish one more year. Daniel and his fiancé, Kristen, would be staying with us for a couple months in Florida. Daniel had taken a job with NMFS in Juneau, Alaska, and they would be driving from Florida to Alaska in the spring after he finished his work with the Southeast Fisheries Science Center in March. We looked forward to seeing them, even though our condo would be cramped with four people and two pet birds. By

Covid halibut! *Left to right*, Eric Graham, Matt Lavigne, and Capt. Dave. They are back! Sixty pounds for sale fresh at the Yankee Coop, April 2020.

the time we reached Florida, the news was filled with stories about a strange, new, contagious virus that had originated in China and was beginning to be seen in the United States. Dan and Kristen had planned on staying with us for the month of February, but the Canadian border closed, and they were stranded. By early March, it appeared the entire country might close, and, worse, Massachusetts planned to close all non-essential businesses by the third week of March. A quick call to the boatyard storing the *Ellen Diane* revealed they were considered non-essential. If I was going to put the boat back in the water, it would have to be by March 21. Ellen and I packed up the birds and ran the Covid gauntlet back to New Hampshire during the second week of March. The coop could still move haddock, but lobster and flounder sales were difficult. I launched the boat, and, having been declared essential food producers, my crew and I resumed fishing.

We had two problems: keeping ourselves away from other people and dealing with observers. The government stopped showing up for work, but fishermen were told they still had to take observers. Fishermen were incredulous. Observers were like traveling Typhoid Marys going on different boats every day. A few angry letters flew off to NMFS and Congress. The day before my first trip, in a rare show of common sense, the observer program was suspended. We tried to exclusively target haddock, much of which we could sell in the coop fish market and a few lobsters. Unloading the boat became surreal. Masks and six-foot distancing were the latest edicts. People would be lined up in a snaking line across the coop parking lot to enter our fish market one at a time to buy our fish. People sometimes stood at the railing along the pier shouting down words of encouragement, thanking us for risking our lives to feed them. This was surreal; we were finally valued for what we had always done, feeding America. NMFS may have thought we were crooks and cheats, ENGOs thought we were rapists of the sea, but to the American people, we were valued members of the community making sure they had fresh food. I am glad I fished one more year. The vindication of my value was emotionally satisfying. While the fishing remained strong, the chances of retirement evaporated. Fishing may have been essential, but all the ancillary actions that surround fishing were not. I was stuck until the country reopened.

By December, the first vaccines became available, and I listed the boat with a broker. The coop was weathering Covid, and Dan and Kristen had finally made it to Alaska, albeit not in the manner they had planned. Their dream of visiting many of the nation's national parks along the way largely evaporated since most were closed. Further, Dan had a letter stating he was essential personnel to allow him to enter Canada, but Kristen was still a PhD student and not essential. Canada refused her entry and informed Dan he could not leave his vehicle during the seventeen-hundred-mile drive to the ferry to Juneau. Luckily Kristen could stay with her parents in Washington State, but Daniel was forced to make the drive without ever leaving his vehicle. Once Dan had established an address, Kristen could fly to Juneau to join him.

Our latest cod genetic work was published, and now no one could say the cod stock boundaries were accurate. This was my second vindication of the year. I had spent twenty-three years trying to convince the scientific establishment that the stock boundaries were incorrect. Now I am reviewing work of a stock boundary working group to offer advice on new stock boundaries. Many of life's loose threads seemed to be getting tied up. Duly vaccinated and boosted, I was not particularly concerned by Covid, and apparently the government was not either. While diligently working from their basements, the observer mandate was reinstated, even though the vaccine had not been approved for young people. My crewman's child and his parents, who babysat him, soon tested positive. The crew continued to test negative, but just the mention of a close contact being positive put the *Ellen Diane* on the do-not-deploy list for forty days. With no observers, spring fishing was nice and peaceful. People were inquiring about the boat, but none wanted to buy the permit because no one would fish in federal waters with 100 percent observer coverage. All the interested buyers planned to fish in state-water-only fisheries in southern New England and the mid-Atlantic. The states had no observer or VMS requirements and did have much less paperwork. Also, all the men who visited to inspect the boat were late middle age. No young people were looking. These two facts should give federal fishery management great cause for concern. People today expect to be treated with dignity and respect. They simply will not put up with being treated like third-class citizens. Absent massive restructuring and simplifying of management, this country will be unable to feed itself from its own waters in a short period of time.

By mid-fall, the government released a new crop of virtually trained observers on the fishermen. Government employees still were not allowed to meet people, so these young trainees had never been given any on-the-water field testing whatsoever. I had one young lady arrive who stated she had never been on a boat before. When I asked if she meant commercial boat, she replied, "No I have never set foot on a boat until this morning." One of my crewmen overheard this and remarked dryly, "Well, this should be an interesting day!" Another woman was seasick before we reached the mouth of the harbor. She rolled back and forth on the wheelhouse deck all day moaning. A man lay in the bunk all day seasick but rallied to observe the last tow. Despite being told where to stand for his own safety, he blundered on deck during haul back, hitting and activating the starboard winch control, nearly ripping the winch out of the deck. I wrote letters and filed complaints, but it was a waste of ink. We had a government run by idiots that would never admit they were wrong.

The day before Thanksgiving, I received a call from my doctor's office. I was tentatively diagnosed with prostate cancer and would need some invasive tests to confirm the seriousness of the situation. Definitely a buzzkill for Thanksgiving. About the same time, I had a serious inquiry about the boat. The man wished to accompany me for a fishing trip. We finally found a day in December with no observer (observers will not go if there are more people than the life raft is rated for). He accompanied me on a successful trip, asked numerous questions, and said he would make an offer when he received the money from the sale of his current boat, which was under contract. I let him know I planned to haul the *Ellen Diane* out of the water in mid-December, and it would be stored until one of us put it back in the water. I went home and told my wife and called my son with the news. Ellen was very upset and so was Dan. I think they both thought I was not serious about getting out, but the cancer diagnosis made it necessary. Eric, always the practical man, thought this was a good idea. Daniel announced he was flying home to make one last trip on his home away from home for fourteen years.

By the time Daniel got a flight, the boatyard had given me a haul out date of December 14. He arrived on the twelfth. I was assigned an observer for the thirteenth. I called the observer provider company, knowing they take more trips than they have people to send on the boats, and told the supervisor the situation. I received the usual speech about mandatory coverage and replied you call me back at 8:30 if you cannot place all your people on vessels and I will take one, but the trip will be canceled at 4:30 tomorrow morning. My son had traveled for two days to make this trip and one way or another he was going out, even if it was only to the next harbor south for the haul out. The phone rang at 8:30 and Ellen answered. She wanted to insert herself in case I still had to take someone. She flashed a relieved smile and told me I was waived.

Early the next morning, the weather was beautiful, with light winds and temperatures above freezing, but the forecast was ugly. A low was forming right over us, and wind would increase rapidly out of the northeast in the mid-day period. My hope was to make two tows and depart the minute the wind picked up. Daniel could take some pictures and tell some stories to the crew about his years at sea. The first tow went according to plan, except we had a cut off lobster trap in the codend, which squared up the diamond mesh and most of the lobsters and haddock escaped. We set out again, and soon I could feel a surge starting from the northeast. I checked the buoy reports off central Maine. They were reporting northeast winds near thirty miles an hour and snow. I hoped to make a long tow running the edge of the bottom for lobsters and then towing to the northeast as far as possible to get a better angle for the ride home. I towed to the south, picked up the edge of the bottom, and followed it back to the northwest. I passed a wreck and turned to tow to the northeast. As I came out of the turn, the boat slowed down and started pulling to the south. Nuts, we probably picked up more cutoff traps. Now the weather was getting nasty. As we hauled back, it became apparent we had something heavy. The boat was traveling straight sideways towards the wreck. I finally got the trawl doors up; one was hanging straight down with a lot of strain but the other one was hanging normal. We must have something on the starboard wing end. My assumption was we had an old submarine detection cable left over from World War II that was probably hung in the wreck. If I could get it up, we could cut it with a grinder. I put the boat in gear and tried to pull the unknown object. The boat jerked and tugged as whatever it was probably slid around the net, but we were not moving. With the rapidly deteriorating weather, this was not a good situation. I was struck by the irony of the fact that I had never lost a net in forty years but might do so today. My crew had never experienced this situation, but Dan had, and stepped in to assume control. They managed to get the net and ground gear disconnected from the doors and onto the net reel. The reel did not have as much hydraulic power as the winches, but it turned, slowly winding up the ground gear. I could not use the engine to apply pressure to the unknown object at this point as the chain that turned the reel might break. Just after getting half the ground gear on the reel, there were a series of violent motions, alternately allowing the stern to rise and then nearly pulling it underwater. Whatever this was it was sliding and periodically hanging up on the net. The boat was bouncing wildly from the wave action, but the port side of the net was visible and the codend was on the surface with the wave action shaking out the catch. About this time there was an extra-large wave which pulled the starboard stern underwater. I ordered Dan to reverse the reel, but as he reached for the handle, the stern shot up and we took off. Whatever it was fell back to the bottom. Daniel and the crew wound the net on the reel and dumped the catch on deck.

Common sense, at this point, would have been to return to port. The tow had about two hundred pounds of cod, a couple big pollack, and about thirty lobsters. I decided to set out again and tow to the northeast. Daniel looked at me and said "You really are an idiot!" This was an inside joke, but my crew did not know it. I explained, while we were setting out, that Daniel used to say it when he wanted an observer to stay out of his way. People figured there was a family rift in progress and a prudent course of action would be to keep a low profile. We towed about forty-five minutes to the northeast, then turned to the north. By now, the wind was well above thirty knots and towing to the north gave me an idea of what the ride home would be like. The answer was: not good. I went about a mile to the north before gradually turning to tow downwind to the southwest. Once the gear was straight behind us, I began to haul back. There is simply a point at which the rewards will never outweigh the risks and we had reached that point. The sea had not caught up with the wind, but it would soon. The *Ellen Diane* traveled about a mile and a half while retrieving the gear. We garnered another twenty lobsters and half a box of cod for our efforts. I told the crew to sit on deck on the bin boards to work and that waves would steadily break over the starboard rail, so make sure the scuppers are clear so the boarding sea can wash back out. The course home was 320 degrees for thirteen miles, ordinarily an hour-and-a-half ride, but today probably twice that. I brought the engine up to about half speed and water poured over the starboard bow and right over the roof. After about a mile, I had to start pulling the throttle back on every wave to keep the boat from becoming airborne. Daniel had been shooting some video with his phone, but now even he looked concerned. The crew was back in the wheelhouse trying to sit on the day bunk. They soon ended up on the deck. Every wave was an adventure. Dan was wedged in the focsle companionway trying to make light of our predicament, saying this was just like coming back from Middle Bank years ago. I felt alive, almost invigorated. I had spent my entire life acquiring the skills to deal with this situation and this seemed like

A solemn goodbye as the *Ellen Diane* is hauled into the Merri-Mar Yacht Yard, Newburyport, MA, for the last time, 2021.

a fitting end to a life at sea. Two and a half hours later, we entered the harbor, cold, wet, tired, bruised, and exhilarated. We had beat nature at her own game. After unloading, I dropped the crew at the Hampton Pier and thanked them for their service. Then Dan and I put the *Ellen Diane* on the mooring. I did not know it then, but that would be my last skiff ride in to the dock.

The wind continued to blow hard all night, much harder than forecast. I had doubts we could get across the bar at the entrance to the Merrimack River but decided to take a ride to find out. Daniel was still sore from the day before and was muttering about life on land making him soft. The ride out of Hampton Harbor was not encouraging. We idled through several breakers, and having reached deeper water, encountered a ten-foot roll from the northeast. On the way down the coast, I explained to Dan what I expected to encounter and what his responsibilities would be. Basically, waves seem to come in series with three larger waves followed by a relative calm period followed by seven larger waves followed by a longer relatively calm period. As with all things in nature, these were generalities, not rules. The buoyed channel in the Merrimack follows the river current out to the southeast. The entire buoyed channel is shallow, less than four-feet deep at low tide. There are sand bars on each side, but there is deep

The last stand. Dan and David leaving for the boatyard. Their last time on the deck of the *Ellen Diane*. December 2021.

water just inside the north jetty. On a day like today, the buoyed channel would be unpassable, but with correct timing you could run the north bar and duck inside the jetty . . . if you were lucky. As we got close, I could see breakers twelve-to-fifteen-feet high on the north bar, but they were not continuous. I had Dan put a couple survival suits on deck and lock the pilothouse door open. If things went horribly wrong, the plan was to jump out the door and grab a suit. I idled at the edge of the bar for a while to time the waves. The waves travel faster than the boat. The goal was to tuck in behind the last big breaker and go as far as we could before the next set caught up to us. Then idle back, let the wave pick us up, and then try to ride just behind the crest like an oversized surfboard. What we absolutely did not want was to get thrown sideways on the face of the wave and get rolled over. Dan's job was to tell me, once I committed to crossing, when a wave broke on the eastern edge of the bar and tell me again when it was almost to our stern, so I could pull the throttle back. After seven waves had crested in front of us, I tucked us in the frothy backside and opened the throttle. I kept waiting for Dan's warning, but it never came. We rode cleanly in behind the breakers, passing the end of the north jetty into the channel. A half hour later, we were entering the travel lift five miles upriver.

Once the travel lift was correctly positioned, I shut down the engine, shut off the batteries, and climbed with Dan onto the travel lift support structure. The lift engine strained, and the *Ellen Diane* rose from the water. Ellen was in the yard lot with a camera and tears in her eyes. Daniel was taking cell phone pictures and trying to appear nonchalant, but soon came over to give us both a tear-stained hug. We all knew an era was ending. A week later, I received an emailed buy-sell agreement. I would not see the *Ellen Diane* again.

CHAPTER 35

What Makes the Players Tick?

Since fishery management is a common thread pretty much throughout my life, I though it useful to include an analysis of the major players and offer a way forward. Most of my comments are in the context of groundfish management, which is the most contentious issue in New England. There are two overarching thoughts to keep in mind. First, management manages people, not nature. Second, there is no one right way to successfully manage a fishery. Everybody involved in this process gets hung up on these two simple points. Many people want fisheries to be managed on specific timelines; nature has other ideas. Also, various factions back one "right" solution, essentially using political will to drown out competing ideas. I have my own views, based on a lifetime in the fishery, which I will offer as a conclusion, but I am under no illusion that my ideas are the only successful way forward. For readers to understand the complexities involved, I think an unsparing analysis of the major players in the management process is necessary. I am sure many will take exception to my views, but I hope they will at least think about what is written before getting agitated.

Fishermen

Fishermen have been portrayed in a negative light throughout my life. While most of this negativity is unwarranted, fishermen do share some of the blame for management failures. Fishermen are not homogenous, but management is, and that is the root of the problem. Fishermen can broadly be broken into two groups: small to medium vessels which dot harbors throughout New England, and trip boats, that is, large vessels over seventy-five-feet long which make 7 to 10-day-long trips and which are confined to half a dozen large, deep-water ports. Boats from the smaller ports are mostly owner-operated and fishing is a way of life. Large vessels, for the most part, are corporate owned and the operator is a hired hand. Owner-operators, whether large or small, have a vested interest in sustainable fisheries. After all, no fish, no fishermen. Corporate interest is often centered on acquiring the largest share possible of the fish and making money for shareholders. As one lobbyist for a contingent of large vessels told me years ago after too much red wine, "We want to be filthy, stinking rich!" This group, unfortunately, controlled council politics in the first twenty years after the two-hundred-mile limit and made numerous poor, short-sighted decisions. There is only one way to get rich quick in fishing and that is very slowly after much hard work. If you are only interested in money, become a stock day trader. More recently, management has pushed consolidation. This too is short-sighted. While trip boats land most of the poundage, dayboats fill in the gaps between trip landings. Remove too many dayboats and the fish distribution system will collapse. Finally, there must be a discussion about criminality in fishing. My personal experience is that fishermen are no different than society in general. I do not believe the percentage of crooked fishermen is any greater than more esteemed positions such as doctors, lawyers, or accountants. Fishermen want these people removed more than anyone else. The states have more effective enforcement than the federal government. Why? State enforcement is on the docks, almost daily talking with fishermen and cultivating friendships. This in turn leads to information on potential illegal activity. Federal enforcement is never on the docks for friendly chats and spends most of its time trying to make large cases with multiple counts. This may make for big headlines, but it undermines their ability to remove problem people before they reach epidemic proportion. Last, I think long-term fishermen suffer from management burnout. For forty-five years, they have been promised short-term pain for long-term gain, but all they have received is more short-term pain.

New England Fishery Management Council

Discussion of the council must be broken out into two categories: council members and staff. To me, staff is the more important of the two groups, because without the staff, the council could not function. Council staff are very hard-working, knowledgeable people who are in a very difficult job. They perform analysis of actions based on council motions, but they also must satisfy a host of federal mandates, including: the National Environmental Policy Act (NEPA), Regulatory Flexibility Act, and Administrative Procedures Act, as well as a host of executive orders and ancillary non-fishing acts such as the

Endangered Species Act. This often leaves the staff to defend their analysis to both the council and NMFS. The council staff is assisted by two groups: the Plan Development team (PDT) and the Science and Statistical committee (SSC). The PDT is made up of NMFS employees, state employees, and occasional academic members. Here, there is unevenness of talent and ambition. The result often is that people with stronger viewpoints have greater say. As a result, analysis may in fact be nothing more than one person's opinion and that opinion is probably more political science than biology. The SSC's function is to review management options to see if they meet scientific objectives. Science operates on consensus, but I have seen a single strong-willed scientist browbeat the entire committee into backing their view. Again, this has the potential for people's political views to be passed off as biology. Both of these groups would function better and provide more useful advice if people remained neutral and kept their personal beliefs out of the discussions.

The council itself is a quintessential political body where a majority of people present determine the outcome of votes. The council is composed of five state directors of the states' natural resource departments, which have different names in each of the coastal New England states, plus the NMFS regional administrator. These people do not have term limits and serve as long as they hold their position within government. In addition, each state has an obligatory seat filled by a resident of the state who is "knowledgeable in the fisheries," as well as seven at-large seats. These are all appointed, with a maximum of three, three-year consecutive terms. ENGOs accuse the council of being dominated by fishermen, but I cannot remember a time when active fishermen ever held a voting majority. If anything, one of the council's biggest weaknesses is a lack of active fishermen and their fine-scale ecological knowledge. Another weakness is a lack of biological background on the part of many members. This lack of biological knowledge leaves much of the council unlikely to question any scientific information it receives. Finally, some council members never speak or go on the record. Therefore, the public has no idea what their thought process is or if they even have one. You cannot screen for that in a first-term appointment, but there is certainly no reason for reappointment. The council is authorized under Magnuson, and as an old adage goes, it's the worst management system ever invented, . . . except for all others. Make no mistake, the council has made some colossally bad decisions over the years, but, in my opinion, most of those decisions were made because the council had too little time to reason through a decision (think thirty pounds of cod made at midnight) or too much outside political pressure from ENGOs and or Washington (think sector management).

NOAA (NMFS) Regional Office

The regional office, overseen by the regional administrator (RA), is the bureaucratic arm of the National Marine Fishery Service (NMFS). NMFS, in turn, is one of several agencies within the National Ocean and Atmospheric Administration (NOAA), which is itself part of the Department of Commerce. In my lifetime, it has grown from a double-wide trailer on the state pier in Gloucester to a four-story building housing hundreds of employees. In the same period, the number of active boats in fisheries it manages has probably shrunk by 80 percent. These figures pretty much tell you why it has problems, as well as why government, in general, needs a massive overhaul. On an individual basis, the employees are dedicated and helpful, but somehow the agency fails on a collective basis. The face it presents to the public is apparently different than what occurs behind the scenes. The agency suffers from mission creep. Its central function is to sustainably feed the American public. To that end, this part of the mission should be moved from the Department of Commerce to Agriculture, which has experience in sustainably feeding America and in working with small entities. Commerce is an agency dealing with business and bean counting and that approach bleeds into NMFS, which collects all manner of unnecessary information. Other NMFS functions, such as protected species and habitat restoration, belong more properly in the Department of the Interior. Many of the agency's problems stem from behind-the-scenes bureaucratic squabbles over which function is more important and lead to fishery-management plans suffering paralysis by analysis. Washington is also to blame, with regional administrators often carrying the water for the career bureaucracy in DC, which operates largely outside of whichever party holds the presidency. The agency also does not delegate authority well. From the public's perception, employees can seldom make an on-the-spot decision. Branch chiefs and legal teams must review even the simplest management decisions. So how can this situation change? The only thing I can think of is to bring in RAs who have medium-size-business management backgrounds and give them a mandate to streamline the agency. Departments should have to justify their information collection and management decision tree. Both Congress and the executive branch could do their part by demanding this as part of the annual appropriations process. As far as fisheries management, the agency should set an overarching goal of sustainably feeding the American public.

Northeast Fisheries Science Center (NEFSC)

While technically under the NMFS regional office, the science center operates on a parallel track as a co-equal. While it suffers from many of the same issues as the regional office, the center has an additional problem. It employs mostly highly specialized and trained scientists. Scientists are notoriously poor at being managed. Their natural scientific curiosity makes them likely to be easily distracted into investigating whatever new scientific question has caught their eye. Scientists also love to collect and analyze data and can become overwhelmed in that pursuit. The center's most important job is to collect and analyze data necessary to produce stock assessments. Several departments feed into that central function, such as trawl surveys, cooperative research, the observer program, and ecosystem and oceanographic analysis, as well as economic and social analysis. In short, it is a large and vital part of fisheries management. So why has such a vital program been the source of so much controversy? Again, on an individual basis the people who work here are friendly, engaged, and very bright, but collectively, like many very bright people, it sometimes succumbs to institutional arrogance. By that, I mean, we are always right, ergo anyone who questions us must be wrong. Population dynamics people have too fiercely defended the status quo at times. This has resulted in newer, improved science being sidelined unnecessarily by a handful of people who either cannot or will not admit science evolves.

While I personally believe our scientists should spend several weeks per year observing the fishery they assess on fishing boats for the reason of collecting their own biological data to insure accuracy while interacting with fishermen and as a means of ground truthing their assessments, I recognize this will probably never occur. That means third-party observers and all the attendant problems. This is one program that needs a major overhaul. It may even need an inspector general probe to find out both the true relationship between the providers and

NMFS and how so much money can be spent to collect useless information totally unrelated to biological necessity and to produce such a large percentage of people that are unfit for sea duty. How the program morphed from a simple, lean, biologically useful program into a bloated, bureaucratic nightmare still astounds and saddens me. I have worked with a small number of people who did the job well and enjoyed the work. Instead of being rewarded, they were hounded by enforcement about false fraternization claims until all of them quit. This situation speaks to a need for management oversight at a high level.

On a bright note, the science center seems to be embracing cooperative research with fishermen as a cost-effective way to improve science. I have read several papers recently endorsing both cooperative research and more incorporation of fishermen's fine-scale ecological knowledge into the fishery science process. This bodes well for the future.

Congress

Congress has one of the lowest approval ratings of any group of people in this country, but I must defend them, at least those from New Hampshire. From the early 1990s, I have worked with New Hampshire's congressional delegation

Senator Shaheen addressing fishermen and citizens about the dredging of Hampton Harbor. Photo courtesy of Senator Shaheen's office, 2018.

from both sides of the aisle. I have found them knowledgeable, engaged, and apolitical when it comes to fisheries. I have answered hundreds of questions over those years and can say not once was a question framed in a political context. I have also found members of both parties to be watchdogs of the taxpayer's purse and engaged in finding cost-effective solutions to fishery management. I think readers should also know that all congressional people rely heavily on their staff. No person could possibly possess the encyclopedic knowledge that would be required on every issue Congress deals with. The staff work incredible hours and produce detailed briefings on every topic imaginable. They are some of the brightest people the public will never know. I would like to think New Hampshire is the rule, but perhaps it is the exception. I attended a meeting where Ted Stevens from Alaska railed against Olympia Snowe from Maine and refused to vote for anything she supported. The Maine congressional delegation would not meet with anyone from outside the state unless accompanied by a resident and a senator from Massachusetts had staff request a minimum two-hundred-dollar donation before meeting with the senator. I would like to think these were exceptions to the rule, but who knows? Magnuson has some faults,

Left to right, Captain Jayson Driscoll, Ellen Goethel, and Senator Shaheen discussing fisheries at the Yankee Fishermen's Co-operative. Photo courtesy of Senator Shaheen's office, 2017.

and I will work with my congressional delegation to see that they are repaired in the next reauthorization.

ENGOs

These groups are difficult to write about, both because of their sheer numbers and the fact that they have such a disproportionate effect on fishery management outcomes. They do, however, share some common traits that need discussion. First, they are big businesses that pay no taxes. Many of these groups rake in millions of dollars with a very simple business model. They sell themselves as the saviors of the planet and the only true arbiters of truth. Next, they pick an animal, preferably a charismatic one with two eyes facing forward and a photographic personality and create an ad campaign to save it. These campaigns should be studied in business schools because they have shown the power of consistent messaging and the ability to raise vast sums of money. Glossy mailings and multimedia events are then created to promote the organization, denigrate anyone or anything that interacts with their chosen species, and finally sue government entities if they do not get their way. Meanwhile, they constantly roam the halls of congress "educating" members in their causes. They are not required to wear the scarlet *L* of lobbyists, because they are nonprofits and nonprofits can only educate, not lobby, but, in point of fact, lobbying is exactly what they do. When you educate, you tell the whole story; when you lobby, you tell the part of the story you wish to sell. I have not seen an ENGO tell the whole story in my entire life. For that reason alone, I believe they violate the tax codes and should have their nonprofit status revoked. Unfortunately for the planet, they employ far more lawyers than biologists. The result is the creation of huge imbalances in the ecosystem as they protect one species to the exclusion of all others. This results in bad outcomes for both the protected species and other species they interact with. Consider seals, whose population is well above a sustainable carrying capacity. Seals carry two parasites which kill the animals and a variation of the bird flu that could jump to humans. Liver parasites and brain worm spread through density-dependent interaction. Thus, the bigger the seal colonies, the more likely for catastrophic thinning of the herds. Bird flu is never discussed publicly in relation to seals. The public is just told they violate the Marine Mammal Protection Act if they get too close. Why? Because the save the seal faction of NMFS and the ENGOs are terrified people will turn against the animals if they learn their pathological possibilities.

So, what do all these lawyers use for legal arguments to convince judges and the world they are right and everyone else is wrong or misguided? In fisheries, the go-to argument is tragedy of the commons. This doctrine was put forward by Garrett Hardin at the University of California fifty years ago. Basically, he stated that the commons would be overgrazed, or, in the case of the ocean, overfished, as humans raced to beat their neighbors to the biggest share of the pie. There are two problems. First, Hardin was a racist and his essay was a convenient argument for white supremacy, and second, it was factually inaccurate. Don't believe me; here are some quotes from *Scientific American* in 2019: "It's hard to overstate Hardin's impact on modern environmentalism . . . It still gets republished in prominent environmental anthologies . . . But here are some inconvenient truths: Hardin was a racist, eugenicist, nativist and Islamophobe." The article goes on to debunk his arguments and point out the commons were quite well managed by local institutions. Remember, fishermen in numerous New England fisheries had ways of managing the commons prior to federal one-size-fits-all management. Think the prohibition on night fishing in Ipswich Bay or the lobster gangs of Maine.

The second major tenet of arguments is the precautionary principle. This is a much more convoluted argument centering around shifting the burden of proof to the proponents of an activity. For fishermen or a government defending their quotas, this means proving the unprovable. The only quota that can be proven risk-free is zero. Thus, the goal of ENGO lawyers is to get words like overfished and overfishing and extinction in their opening arguments as often as possible. No judge wants to have extinction on their conscience, so the ENGOs often pull out a victory without proving their desired outcome factually. This scenario is currently playing out with right whales and the lobster-trap fishery. The ENGOs cannot prove that continued use of vertical lines will lead to the whale's extinction, but the fishery cannot prove that ropeless gear will not save the whales. Ship strikes and genetic inbreeding may doom the whales anyway, but under the precautionary principle, a judge has ordered NMFS to act. Finally, government gets due deference from the judiciary under a legal precedent called the Chevron deference. Basically, if a statute is unclear or ambiguous, the agency gets wide latitude to interpret the statute and provide proof as the recognized expert. Thus, suing NMFS over its own interpretation of its guiding statutes is nearly impossible. Judges will freely admit they do not know filet of sole from the sole on the bottom of their shoes, and therefore NMFS must be right. I often wonder how we would eat if agricultural policy was written the same as fishery policy. Hey, earthworms have rights, too!

CHAPTER 36

Dictator for Life

Given the convoluted and intertwined relationships of the principal players in fishery management, I am often asked how the process can be improved. The short answer is: probably it cannot be made better. No one surrenders power voluntarily, and there are too many power bases in the current regime. Assuming for a moment, that I was made fish dictator for life, a scary thought, I would request the following changes.

Magnuson has some obvious shortcomings that need rectifying and some changes to make what I propose legal. Number one: stop requesting that all species be brought to maximum sustainable yield simultaneously. This is biologically impossible, as many of these species compete in the ecosystem. Next, revise definitions of overfished and overfishing to deal with species that are having issues unrelated to fishing. Ecosystem shifts, regimes change, and other environmental challenges need a different definition. It should be common sense that a species cannot be overfished or have overfishing occurring if there is no fishing. Gulf of Maine shrimp is the classic example. No gear interacts with it, and there has been no fishing for more than the animal's lifespan, yet it is still listed as "overfished" because of word definitions. Come up with new terminology to deal with environmental issues that occur beyond management control. Eliminate the requirement that species be managed on a single-species basis. Some animals, like scallops or lobsters, can still be managed the traditional way, but animals that swim together, like groundfish, should be managed on a regional ecosystem basis. The sum of the single species parts is always greater than the ecosystem whole. We must find a way to use basket-total allowable catches and stop worrying about single species. Why? Because the basket represents what is there. Fish do not swim singly without interacting with other fish.

Finally, reference points should shift with ecosystem conditions. Warm-water regimes produce fewer cold-water fish than cold-water regimes. No matter how much NMFS and the ENGOs want to return to the cold-water regime of the 1960s, all current indications are it will not happen. Deal with what is, not what you would like. Also, find a way to issue permits to people based on what lives at their doorstep. Many fishermen in the Gulf of Maine do not have permits for fish like scup, butterfish, squid, fluke, and black sea bass, which have moved to this area as waters have warmed. This leads to a large, unnecessary discard issue, as fishermen can only retain what they have permits for.

Once these legal changes have been implemented, management needs to change. Sector management, Jane Lubchenco's baby, needs to be eliminated. A system based on greed cannot be expected to foster anything but more greed. Think of the example of Carlos Rafael cited earlier. Carlos did not care about the long-term sustainability of the fishery; he only wanted more money and more fish for Carlos. That is greed, and this system fostered that behavior. The system is prohibitively expensive to administer and benefits neither fish or fishermen. I would replace this with a two-tiered zone system. Offshore is anything beyond twenty-five miles and inshore is within twenty-five miles. Figure out how many vessels you need up and down the coast to harvest the fish in these zones on a sustainable basis by gear type, cancel inactive permits, and eliminate excess active permits and vessels. Buy out excess capacity from the two zones. In the event voluntary capacity reduction does not achieve the desired number, place the remaining vessels in a lottery and involuntarily retire the losers. Apply some broad controls, such as mesh size and spawning closures, and let the remaining boats fish. Shrink the NMFS bureaucracy and eliminate unnecessary programs like at-sea monitors, and use the money saved to fund the buyout. Eliminating unnecessary jobs may not provide sufficient capital for the buyout. In that event, Congress should appropriate the money. A one-time expenditure will be far cheaper in the long run than continuing to fund the current system. The council's new charge would be monitoring the fishery and setting quotas for other fisheries, mandating gear improvements as they become available, and overseeing cooperative research. In the event the multispecies total allowable catch is exceeded two years in a row, an appropriate further reduction in vessels would be initiated. The goal is to have a fishery that operates sustainably without enormous constraints and inefficiencies that require large and continuing bureaucratic oversight.

I am sure by now some readers are either laughing out loud or plotting how to put arsenic in my drink. Why? Well, I have gored every major player's

The F/V *Ellen Diane* docked at the Isles of Shoals Marine Lab, Appledore, ME, 2009.

ox. Inactive fishermen will be mad their permits are canceled. Multinational corporations that are waiting to buy up the fishery will be upset their plans have been thwarted. If vessels need to be involuntarily retired, those fishermen will be mad about losing their occupation. The bureaucracy will be upset, but since they cannot be fired, their aggravation will be confined to burrowing into other government departments. The ENGOs will be upset, because the lawsuit mill will be closed. Maybe their celebrity sponsors will keep them on retainer. The only group that benefits is the council and council staff, who should have a much easier job managing the fishery going forward.

Why did I save this for the end? If I had presented these ideas in the mid-1980s to readers, they would have thought them far too radical. I felt you had to see the progression from where we started to where we are to have anyone give this idea serious consideration. For what it's worth, I did discuss this with council staffers in the late 1980s. I was assured that management would control groundfish overfishing, and the short-term pain would not last much longer.

US Fisheries Management Structures

There are a lot of acronyms and other references to state, federal, and even international organizations this book. Imagine what it's like being a fisherman and having to deal with them all, let alone keep them straight. Hopefully these brief inclusions will provide a reasonable introduction to the industry.

NOAA Management Regime

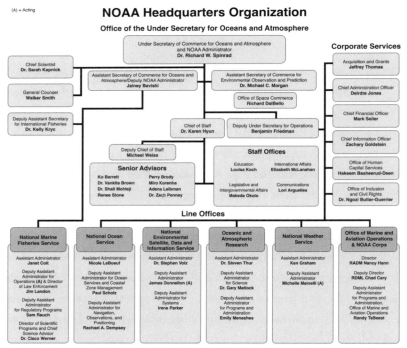

This is the organizational chart for NOAA headquarters leadership and senior staff as of February 2, 2023. This is included to give an idea of the complexity involved. The eight regional councils report their recommendations to the National Marine Fisheries Service (NMFS), colloquially known as NOAA Fisheries, shown in the bottom left. From https://www.noaa.gov/about /organization/noaa-organization-chart.

U.S. Regional Fisheries Management Councils

The **Magnuson-Stevens Fishery Conservation and Management Act (MSA)** is the main law that governs fishing in U.S. federal waters, ranging from 3 to 200 miles offshore. First passed in 1976, the MSA established a 200-mile Exclusive Economic Zone (EEZ) and created eight regional fishery management councils to manage our nation's marine fishery resources. This unprecedented management system gives fishery managers the flexibility to use local level input to develop management strategies appropriate for each region's unique fisheries, challenges, and opportunities.

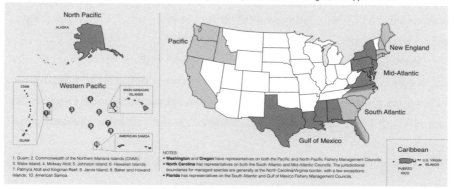

Image adapted from the U.S. Regional Fishery Management Councils PDF "2019-05-15_RFMC-Overview_UPDATED_FINAL" (http://www.fisherycouncils.org/).

New England Fishery Management Council (NEFMC): Organizational Structure

The NEFMC and seven other councils were set up by the Magnusson-Stevens Act. These councils are not part of government per se, and neither council staff nor council members are federal employees. There are eighteen members of the New England council. Each coastal state has one state employee on the council. In addition, each state must fill an obligatory seat with a person "knowledgeable of the fisheries" who is a resident of that state. Additionally, there are seven at-large seats, which, technically, are not representing a specific state. Finally, the regional administrator of the National Marine Fishery Service (NMFS, see NOAA Management Regime, above) is a voting member.

The eight councils are advisory, but NMFS must take their advice unless NMFS's lawyers determine part or all of the advice is not legally defensible. NMFS can reject specific parts of a framework or amendment, or all of one on legal grounds. I am not aware of an entire action being rejected, but several times specific parts of actions were rejected. For example, government agencies are bound by many acts beyond the Magnusson Act. We had a requirement for herring fishermen to pay NMFS for fishery observers rejected because NMFS cannot collect money. All money collected by government must be paid to the treasury. That is why you write your tax checks to US Treasury and not to the IRS. NMFS also has statutory authority to take emergency action if it feels a species of fish is becoming endangered or dropped to critically low levels. This was used in 2014 to cut cod quotas 90 percent after an unscheduled stock assessment, which is discussed in the book.

The following is quoted from the New England Fishery Management Council website description of themselves: https://www.nefmc.org/about/history. Council recommendations are forwarded to the National Marine Fisheries Service, a.k.a. NOAA Fisheries (see NOAA Management Regime, previous page):

Each **Council** has established its own process, broadly outlined in Section 302 of the MSA, to accomplish the work of developing rules that apply to the "managed" fisheries that operate within its areas of responsibility in the US exclusive economic zone. The New England Council relies heavily on its Oversight Committees, Advisory Panels, Plan Development Teams and Scientific and Statistical Committee to develop management actions.

Oversight Committees allow the Council to more efficiently develop alternatives and management measures for consideration and eventual inclusion in a fishery management plan (FMP). Each Council member serves on one or more oversight committees. Committees are generally related to a specific fishery or important management issue, such as habitat or ecosystem-based fisheries management (EBFM).

Committee members develop specific measures that will form the basis of the plan, plan amendment or framework adjustment to an FMP. Oversight committee recommendations are forwarded to the full Council for approval before they are included in any draft or final FMP.

Advisory Panels are made up of members from the fishing industry (from both commercial and recreational sectors), scientists, environmental advocates, and others with knowledge and experience related to fisheries issues. They meet separately or jointly with the relevant oversight committee and provide assistance in developing management plan measures.

Advisors are appointed every three years following a solicitation for candidates. After reviewing applications, the respective oversight committee recommends new or returning advisors. The Council's Executive Committee provides final approval of advisory panel members.

Plan Development Teams or (PDTs) provide an expanded pool of expertise for the purpose of conducting analyses and providing technical information to the Council. The PDTs also help ensure that Council FMPs, amendments, and framework adjustments meet scientific and legal requirements for review and approval.

Oversight committees may ask their PDTs to evaluate management proposals, develop options to meet FMP objectives, or to provide guidance on a variety of scientific, technical or FMP implementation issues. They provide technical analyses concerning species-related information, and develop issue papers, alternatives, and other documents as appropriate.

The NEFMC's Scientific and Statistical Committee provides the Council with ongoing scientific advice for fishery management decisions, including recommendations for acceptable biological catch, preventing overfishing, maximum sustainable yield, achieving rebuilding targets, and considerations related to the social and economic impacts of management measures.

The Council considers the recommendations of its committees, the advice forwarded by its Advisory Panels, and the analyses provided by its Plan Teams, as well as the testimony of affected stakeholders and the public in developing proposed management actions.

About the Author

David Goethel retired from commercial fishing in 2022 and splits his time with his wife, Ellen, and parrots, Huey and Stuart, between Hampton, New Hampshire, and Stuart, Florida. In Stuart, he surf casts for pompano and other species. In Hampton, he fishes recreationally for the wide range of species in the Gulf of Maine. The author remains active in both science and fishery management at both the state and federal level, serving on one council advisory panel and one ASMFC advisory panel. He also serves on two volunteer fishing organizations' boards of directors, helping to promote sustainable fishing practices. David continues to promote cooperative research, scientists and fishermen working on research projects together to aid in solving the many problems that still plague fishery management.